W9-BCY-758

Cleavage

CLEAVAGE

Technology, Controversy, and the Ironies of the Man-Made Breast

NORA
JACOBSON

RUTGERS UNIVERSITY PRESS
New Brunswick, New Jersey, and London

Library of Congress Cataloging-in-Publication Data

Jacobson, Nora, 1964–
Cleavage : technology, controversy, and the ironies of the man-made breast / Nora Jacobson.
p. cm.
Includes bibliographical references and index.
ISBN 0-8135-2714-7 (cloth : alk paper). — ISBN 0-8135-2715-5 (paper)
1. Breast implants—Social aspects. I. Title.
RD539.8.J33 2000
618.1'90592—dc21 99-23192
CIP

British Cataloging-in-Publication data for this book is available from the British Library

Manufactured in the United States of America

Contents

Acknowledgments

In preparing these acknowledgments, I was surprised and gratified to realize the number of people I have to thank.

In its dissertation phase, this work was improved by the comments and suggestions of many faculty at the Johns Hopkins University School of Hygiene and Public Health, especially Margaret Ensminger, Elizabeth Fee, Constance Nathanson, Jacqueline Corn, Carol Weisman, Barbara Curbow, Norman Anderson, and Jeff Johnson. My thanks also to Lavern Riggs and Mary Jo Dale, of the Faculty of Social and Behavioral Sciences, for their support; to the staff of the Interlibrary Loan Department at the William H. Welch Medical Library, for fielding innumerable requests; to Robert M. Goldwyn and Madeleine Mullin, for making the resources of the National Archives of Plastic Surgery available to me during a week-long research visit in the spring of 1995; to Laury Oaks, for the time she devoted to my work; and to Ted Chiappelli, for the weekly telephone calls that kept me company.

The transformation from dissertation to book, a transformation that has been several years in the making, was supported by Joy Perkins Newmann, director of the Mental Health Services Research Training Program at the University of Wisconsin, Madison (NIMH grant #5T32MH14641), where I was a postdoctoral fellow from 1997 to 1999; by Barbara Bowers, my faculty adviser in the program; and by Vivian Littlefield, dean of the School of Nursing at the university.

My Madison research group—Patti Herman, Kristy Ashleman, and Chantal Caron—eased the writing process immeasurably, offering much needed commentary on a regular basis. Amber Ault has been a careful reader and a helpful critic. Bonnie King and A. B. Orlik provided calm technical assistance at important junctures. Many other friends and colleagues, both in Madison and in Baltimore, took a lively interest in the progress of the book.

Barbara Katz Rothman has been the source of much encouragement and good advice.

At Rutgers University Press, I have enjoyed working with Doreen Valentine, Martha Heller, David Myers, Tricia Politi, Brigitte Goldstein, and Victoria Haire.

I owe my parents, Dolly and Harvey Jacobson, more thanks than I can offer here.

Cleavage

INTRODUCTION

Consider, if you will, these two sets of images:

The headline of a newspaper advertisement reads, "Millions of women have chosen implants. You may never have that option." Beneath the headline are photographic portraits of four women, all poised, attractive, well groomed. Each seems to have been interrupted in the midst of speaking. The first woman, Lynne Lofgren, says, "My recovery from breast cancer started with my decision to have reconstructive surgery." She directs her gaze just to the side of the camera, and her chin rests on her hand. Ruth Heidrich looks into the lens: "All I wanted was to reconstruct my body and my life. I don't want the government making my medical decisions." Her gently curled fist brushes against the side of her face. Linda Sains's eyes are narrowed in concentration. Her thumb and index finger are poised by her jaw, gesturing as though to emphasize the importance of her point: "Building self-esteem is a personal issue. As an adult, I want to be able to make my own choice." The fingers of Dotti Nacci's hand are spread across her chest, just below her throat. "Women should take the responsibility to gather the information on breast implants and make their own decisions." She too gazes directly into the camera.

The text below the photographs and their captions argues

against government restrictions on breast implants and urges readers to contact their congressional representatives to demand their support for "one of the most personal decisions a woman can make." A line at the bottom of the page notes that the advertisement is a message from the American Society of Plastic and Reconstructive Surgeons (ASPRS).

Two photographs are placed side by side and set into a page of text. The one on the left is captioned "before." It shows a woman's body, naked, from waist to neck. She is headless, handless, anonymous. Her chest is angular. The photograph on the right is captioned "after." It shows the same woman's body, still naked and anonymous. Her breasts are now larger, full. The text surrounding the photographs recounts the "during," a technical tale of the surgery that effected this transformation.

In the fall of 1991 the ASPRS advertisement appeared in major newspapers across the United States. Days later, the Food and Drug Administration (FDA) was to hold a public hearing about implants before a panel of medical experts. The panel would determine if there was sufficient evidence to prove the safety and effectiveness of the devices, then make a recommendation to the FDA about whether the agency should allow implants to remain on the market. The procedure was familiar, a bureaucratic protocol that had taken place thousands of times since 1976, when Congress had given the FDA jurisdiction over medical devices.

Much more than protocol was at issue this time, however. In fact, the FDA hearing was to be the culmination of months of controversy over the fate of silicone gel breast implants. That controversy had seen a vituperative congressional hearing, splashy stories in the media, and costly courtroom dramas. Women claiming to have been harmed by implants had made public their stories of private suffering. Satisfied implant recipients had countered with their own tales of restoration and improved self-esteem. The FDA, under criticism for being both too zealous and too lax, had shown that it would assert its authority over this issue. The manufacturers of implants, most notably Dow Corning, had mounted a massive public relations campaign designed to increase support for implants. Similarly, plastic surgeons had become activists, giving interviews to the press, lobbying Congress, and soliciting their patients to join the good fight.

The ASPRS-sponsored advertisement highlighted many of the questions at the center of the breast implant controversy. Who were the women who had received implants, and why? How was benefit to be judged, and by whom? What were the risks of implants? How could risks, and benefits, be assessed? Was a history of clinical use enough to prove safety, or must specific scientific data be provided? What was the proper role of government in regulating medical devices? Was the "right to choose" an absolute? Whose breast was it, anyway?

The before-and-after photographs have appeared many times between the late nineteenth century and the present—the time period encompassing the history of breast implants. Their setting is a medical journal, most often one devoted to plastic surgery. The photographs are illustrative, demonstrative. They are tools meant for colleagues, those surgeons who may want to replicate the results.

The questions these photographs raise are either technical or personal (perhaps disturbing or titillating). How was this transformation achieved? Whose body is this? What happened in the moments just before and after the shutter snapped? The photographs promise a story in which the woman is central, but the anonymity of the body—its face and hands cropped away—implies that she is more object than subject. The "I" of the narrative is the plastic surgeon. The "woman in the body" remains silent.[1]

The aim of this book is to ground the questions raised during the controversy in the history of breast implant technology by examining the meanings of the technology over time. To understand the controversy, we will look at how these meanings were constructed by specific actors, including plastic surgeons, implant manufacturers, the FDA, the media, and women themselves.

Technology provides the skeleton of this story. Silicone gel breast implants did not arise de novo. Rather, they were one in a series of technologies dating back to the late nineteenth century. Silicone gel implants followed foam implants, which followed transplantation of fat and derma, which followed paraffin injection. The periods of use of these technologies are not distinct. Some silicone gel implants incorporated foam shells. Modern techniques of reconstructive surgery move tissues from one part of the body to another. Reports of paraffin injections continued into the late 1970s. Each technology, however, had its era of ascendancy and decline. An introduction with great promise was followed by a period of revision suggested by problems of its use. Finally, the technology would lose its dominance and was superseded by the next innovation. This cycle of innovation-modification-disillusion occurred many times in the history of breast implants.

The history of breast implant technology is intertwined with the history of three surgical procedures: augmentation, enlarging a small breast; reconstruction, rebuilding a breast lost to mastectomy; and prophylactic mastectomy, removing healthy breast tissue and benign masses against the threat of future disease. These procedures differ—by indications, by patient population, by rhetoric, and by the extent to which they are accepted—but they also have much in common. The demands of each procedure shaped the cyclical development of breast implant technology. As each procedure gained popularity, the application of the technology became more widespread, provoking more innovation and modification, and introducing more chances for disillusion.

My characterization of breast implants as a technology may surprise readers who are used to thinking of technology as cold metal or electronic circuitry. I use the word as it is defined by Susan Bell, who wrote that technology is "a product or embodiment of human activity" which "contains concepts as well as political, social and economic structures,"[2] and Donna Haraway, who called technologies "instruments for enforcing meanings."[3] That is, technology is a manifestation of society, the historical moment made material. The history of breast implant technology is the story of the construction of many shifting political, social, and economic meanings.

Bell's definition implies that technology reflects society, while Haraway's suggests that it also may shape, or direct, society. The tension between technology as reflection and technology as direction is central to this study. Did plastic surgeons invest their time and resources in breast implants because they were confronted with a demand for the devices? Or did the very existence of breast implants, and the efforts of the surgeons who developed them, promote demand?

Much of the extant research on the meaning of breast implants has focused on the reflective significance of the individual demand for these devices. That is, it has sought to determine just why women want implants and what such desire says about society. The literature has offered two explanations. In the first, the individual desire for implants is a response to the multitude of societal rules and strictures that define what it means to be a woman.[4] In this formulation, breast implants embody the "paradoxes of choice," the irony that our most deeply personal yearnings are but a reflection of the world around us.[5]

Critics of this explanation, however, argue that such analyses fail to give women their due as autonomous agents and turn them into dupes and victims by suggesting that their choices are but mindless capitulations to social pressures.[6] These researchers turn to examining the lived experience of cosmetic surgery in the stories of women who choose it. Their findings suggest that for many women breast implants and other cosmetic surgery become powerful and empowering ways to negotiate new relationships with self and others.[7]

This research on the question of why is both unsatisfying and troubling to me. Its analytic focus on the individual suggests that the individual should be the locus of change. Conversely, its references to society are too unfocused. What is this society? Where is it? It is inchoate, permeating. Because it explains everything, in the end it falls short of explaining anything. Finally, the technology itself gets lost in this approach: in its status as symbol it loses any historical specificity.

My way out of this conundrum is to take a different approach. A focus on the instrumentality of implants (technology as direction) allows me to pay less attention to individual choices and more to the purposes and processes of the specific actors who constructed the various meanings of the technology. Looking at how the meanings of implants were framed, and for what purposes, allows me to explore the ways in which the technology shaped and was shaped by different groups and institutions. Posing the central problem in this way lets me examine parts of the implant story that have not been extensively analyzed before. It draws attention away from the question of why individuals make certain choices. It locates "society" fairly specifically, in the circumstances surrounding the construction of specific meanings.

For example, this approach led me to make a conceptual distinction between desire and need. If desire is the individual internalization of society, need is the legitimized form of desire. While desire may have preceded and helped to drive the development of the breast implant technology (technology reflected desire), something called need was the result of it. The potentialities of the technology, which developed through the cyclical process to which I have alluded, suggested how it could be applied. The existence of the technology necessitated the legitimization of its use; thus technology defined need. Conceptualizing need as one of the meanings of implant technology provides one way to interpret the events of the breast implant controversy, where need was an important—although not heavily publicized—issue. While desire is private, need is public. There is good evidence about need, evidence that is lacking for all but the most recent accounts of desire. Finally, because what is considered legitimate varies across time, need links the technology and its historical location.

Early on, the primary legitimization process was medicalization, or the "process whereby more and more of everyday life has come under medical domination, influence and supervision."[8] The first breast implant technologies—paraffin injections, tissue transplants—were used infrequently. In case reports surgeons offered little justification for their use, focusing instead on the technical aspects of the procedures. When authors did characterize the indication for surgery, they blamed it on women's unhappiness and dissatisfaction with their bodies. With the advent of synthetic sponge implants (circa 1950), and the wider potential for the use of the technology, the indications for use became certain syndromes: "hypomastia" (literally, small breasts, also called hypoplasia and micromastia) in cases of augmentation and "postmastectomy syndrome" in cases of reconstruction. Each diagnosis was understood to be a psychological complex—feelings of inadequate femininity resulting from inadequate anatomy. With prophylactic mastectomy, plastic surgeons tried to construct need out of a physical indication—fibrocystic disease (a catchall term for benign breast lumps and nodularities)—and to establish the subcutaneous mastectomy with immediate reconstruction as primary prevention against breast cancer. As variations in the size and shape of the breast became "diseases," plastic surgeons offered breast implants as treatment. These attempts to legitimize breast implant technology were tied up with plastic surgeons' efforts to establish the legitimacy of their profession.

In the 1960s, the introduction of a new technology—the silicone gel implant—and important social changes—including increased affluence and the rise of the women's movement in America—converged to provide other avenues of legitimization. Increasingly, plastic surgeons promoted cosmetic breast surgery as fully justified for any number of indications: psychological, social, and economic. In the 1970s, vanity, the bugaboo of the profession, was normalized, then glorified. Beautiful breasts became just another commodity. As need was depathologized, the number of potential candidates for surgery increased. As more women sought surgery, plastic surgeons placed new requirements on the technology, demanding an implant that would look and feel soft and natural.

The foreshadowings of the breast implant controversy lay in these attempts to modify the technology. Plastic surgeons and manufacturers, for whom breast implants had become a lucrative and competitive business, rushed to produce a new generation of implants. Like their predecessors, these implants did not undergo any sustained scientific testing before reaching the market. Certain characteristics of the new technology, specifically its thinner envelopes and looser gels, brought new possibilities for things to go wrong. It was very soon after the introduction of these new implants that reports of systemic illness linked to the devices began to appear in the medical literature.

During the same period—the late 1970s and early 1980s—the FDA was granted the authority to regulate medical devices and showed its first interest in breast implants. Implant manufacturers and plastic surgeons were well aware of the potential threat to implants posed by government regulation, but believed that they, as the experts in their use, would play a central role in making policy. At first, it appeared that they were right. Until the late 1980s, the

FDA largely deferred to the plastic surgeons and manufacturers, trusting them to perform their own examinations of the devices. All of this changed in the antiregulatory climate of the early 1990s. As the very existence of the FDA was threatened, breast implants became an issue through which the agency could assert its authority.

For many readers, the most recognizable picture of implants may be the one that emerged during the FDA controversy, when the media portrayed implants largely in terms of risk and harm. Anyone who tuned in to the news at the time remembers the frenzy: silicone victims weeping, corporate spokesmen denying, investigative reporters confronting, plastic surgeons defending. After the public hearings, the FDA placed restrictions on the availability of implants. Because media reports often oversimplified the agency's rationale for its decision, the lasting impression was that the devices had been judged too dangerous for the market. In this atmosphere, breast implants became the exemplar in discussions of the limits of societal protection and the avenues for individual compensation.

What's past is but prologue. The events since 1992 have reconstructed the technology's meaning yet again. A number of epidemiological studies have failed to prove that implants are harmful. Huge amounts of money have been offered to settle class action suits. Plastic surgeons have continued the cycle of technology development, working to produce new and better implants. Increasingly, social critics have promoted the breast implant story as a paradigmatic case of the politicization of science, or "junk science," raising questions about the conditional nature of knowledge and truth and the certification of experts in a diverse society.

The origin of my interest in breast implants lies, appropriately enough, in a women's magazine. In the fall of 1993, I happened to buy the latest issue of *Self,* a glossy magazine devoted to women's health and appearance. Among the articles was one by Joyce Maynard called "After the Silicone" about her recent experience with breast augmentation surgery (or, in glossyspeak, "the journey from 32A to 38C").[9] In contrast to much of the coverage of silicone breast implants then appearing in the popular media, Maynard presented a mostly positive account, acknowledging the possibility of health risks, but emphasizing how breast implants helped her to embody a renewed sense of her own sexuality after the physical and psychological changes wrought by the births of three children and the dissolution of her marriage.

I was struck by the language Maynard used to describe the possible transformative effects of plastic surgery:

> Inhabiting a tense and chilly marriage as I had for some time, I had come to see my breasts as a symbol of my own diminished sense of my attractiveness and sexuality. When I fantasized about plastic surgery, as I did even then, what I was really dreaming of was something unattainable in any physicians office, of course. It was a better feeling about myself.
>
> Some years later, my marriage over, the old dream resurfaced. But this time, when it did, I was no longer looking for breast surgery to "fix" me. . . . I sought out a plastic surgeon at a time in my life when I was in love, in bloom, alive with my own sexuality for the first time in years. . . . I wasn't trying to change my sense of myself by changing my physical form. I simply wanted my body to look, on the outside, the way I was feeling within.[10]

This narrative of desire contained a number of interesting ideas. Maynard suggested that some women do in fact have breast augmentation in order to "fix" their dissatisfaction with their lives—"to change [their sense of self] by changing [their] physical form." She implied that breasts are symbols of maternal nurturance and of female sexuality. Implicit was the assumption that bigger, perkier breasts are better breasts—more feminine, more attractive. (I was also intrigued by the idea that blooming love can be measured by bra size—38C, no less!)

What Maynard's story lacked was (excuse the pun) proportion, a sense of the cultural context in which breasts become symbols and women resort to surgery for social problems. Such a lack was striking, especially given the article's immediate context—contemporaneous journalistic accounts emphasizing the medical, legal, and political controversies surrounding implants—and its placement in a women's magazine engaged in a contradictory struggle over representations of the "ideal" female body. (The text of magazines like *Self* urges women to learn to love the body they have and to demonstrate that love by giving it good food, good exercise, and the proper material accoutrements—breast implants apparently falling into the latter category. But the photographs in these magazines hold up the bodies of fashion models as visual exemplars, bodies perfected by techniques of deception, like airbrushing, and surgical manipulation, like implants themselves.)

The article piqued my sociological imagination.[11] In graduate school at the time, studying public health, I wrote a paper about silicone gel breast implants, analyzing them as the public health problem that they had become. Using an approach based on Joseph Gusfield's model of the social construction of public problems,[12] I posited a series of interrelated problems: the health problem, the suffering problem, the bureaucratic problem, the corporate problem, the technical problem, the legal problem, and the body image problem. I named the various "owners" of these problems and examined their interests in them. For example, I noted the financial and professional investments of the manufacturers and plastic surgeons seeking to ensure the continued availability of implants and also explored how the goal of restricting implant availability was essential to validating the illness experience of the "silicone victims."

I finished that paper with many, and more interesting, questions. My research revealed that there was a huge gap in the literature: a historical account of the development of implants and the eventual controversy surrounding their use. Without such a history, I felt I could never understand the social phenomenon of implants. I made breast implants the subject of my doctoral dissertation, using the social construction of public problems approach I'd taken in that first class paper to look at the history of breast implants. My aim was not to draw authoritative conclusions about the contested realities of implants, but to look at how those realities were constructed, by whom, and for what purposes. Both my approach and the topic were somewhat unorthodox for a student at a quantitatively oriented, very positivist school of public health. I spent a lot of time explaining what I was not doing. I was not doing epidemiology, designing a study to assess the truth of women's claims of harm. I was not doing psychological or behavioral research, seeking to find a model that could explain why women wanted implants. My work was not—horrors!—journalistic in intent; the point was not to muckrake. I sensed some speculation about why I was so interested in implants, and grew accustomed to people glancing at my chest as I talked about my project. Sometimes I wondered myself, in a world of such disparity in health and access to care, why had I chosen to focus on something so seemingly frivolous?

One response to that question is to dispute the charge of frivolity. The recent history of silicone breast implants stands as a case study of the construction of a public health problem. The complex interplay among scientists, lawyers, industry, government, physicians, and patients in the struggle to "own" the problem resembles, and thus can help us to understand, the pattern of conflict seen in many other more mainstream public health issues, like tobacco or automobile safety. As a women's health issue, breast implants illustrate recent critiques of women's health care and policy, critiques that focus both on the abusive excesses of medical treatment and on the lack of attention paid to particular women's health problems.[13] Thus breast implants can teach us about those issues as well.

Social constructionism, the methodological approach used in this book, can expand our understanding of many social problems that involve health and illness. Based on the sociological tradition of symbolic interactionism, social constructionism looks not at the "objective realities" of societies but at the processes through which social actors create their own realities.[14] This perspective situates individual behavior in society, while acknowledging that individual (and collective) acts create society. The conceptual inseparability of the individual and the society, and the focus on the concrete processes that structure their realities, point the way to innovative, creative rethinkings of many public health problems.

The social constructionist approach suggests a method. In researching this book, I collected the documents that trace the history of implants. My sources included medical and scientific journals, textbooks, proceedings of conferences and symposia, instructional films, newspaper and magazine articles, first person narratives, newsletters, transcripts of public hearings, official government reports, and legal documents. I approached these sources with questions about meaning in mind "What is the meaning of breast implants in this text?—then immersed myself in their language, noting and comparing the definitions they contained. I learned more about these definitions while linking them to specific actors and particular circumstances. Finally, I used theory to help me draw connections between my data and generalizable issues of historical and sociological import.

This book is organized both chronologically and thematically. Chapter 1 sets the stage for the analysis to follow by recounting the events of the silicone breast implant controversy of 1991–92. This chapter introduces the major actors in the implant story and examines the meanings of implants that were in the spotlight during this period. Chapters 2 and 3 offer a history of breast implant technology. This history begins in the late nineteenth century with a German surgeon named Vincenz Czerny, and it continues into the present. Topics explored along the way include how the use of artificial materials for breast surgery, initially the subject of dispute within the medical profession, gained widespread acceptance; the evolution of the surgical procedures for which implants were used; how cosmetic breast surgery came to seem commonplace and normal; and shifts in the meaning of the breast itself. Chapter 4 deals with the ways in which the need for breast implants was constructed (and reconstructed) by plastic surgeons as they developed the technology. Chapter 5 looks at the history of conceptualizations of harm caused by silicone. In particular, it examines the creation of silicone disease as a physical manifestation of the perfidy of plastic surgeons and the malfeasance of implant manufacturers. Chapters 6 and 7 are in-depth examinations of two of the major actors in the implant controversy. Chapter 6 offers a brief history of plastic surgery, then shows how that history was played out in the actions and reactions of the profession during the controversy. Similarly, chapter 7 looks at how the history of the FDA, and the political pressures of the Reagan-Bush era, contributed to the agency's position during the controversy. Finally, chapter 8 summarizes the medical and legal events that have taken place since 1992 and explores how the meaning of implants continues to change.

That I claim for this book no authoritative stance—it is, after all, only another construction—brings me once again to the self-imposed charge of frivolity. This time, I embrace it. Breast implants are absurd. Their history is full of dark humor, multiple ironies. This chapter opened with two contrasting sets of images: the articulate advocates of breast implant technology and the stark nakedness of the female bodies upon which that technology has been used. Like all images, they are constructions: of desire, of need. Their juxtaposition epitomizes many of the ironies of breast implants. I see this story through a double lens—the one serious, the other ridiculous. Only a parallax view provides dimension.

1 STAGING BABEL
A Chronicle of the Controversy

I too was a happy patient until the silicone time bomb went off in my body. This is where my uninformed right to choose breast implants has gotten me today. We believe that a woman has a right to choose a safe, tested, and approved product, not a poorly designed experimental product. . . . Now that problems are beginning to surface, listen to the guinea pigs who paid up front to choose, when we were the most vulnerable. . . . When we put our health and lives in doctors' hands, we expect to be told the whole story, not just a slick sales job of cosmetic benefits with a promise that implants would last a lifetime.

—Janet van Winkle, testimony to the FDA, November 12, 1991

Women who choose to have their breasts restored should not be judged by society as vain or even foolish. One who has not experienced breast cancer may intellectually analyze the risks versus benefits and make that assumption. However, I think the logic of another breast cancer friend who had implants explains it best when she says, "I've had chemo, I've had radiation, I have friends around me dying of this disease, but when I get dressed in the morning and I look in the mirror, I am not constantly reminded that I may die."

—Rosemary Locke, testimony to Congress, December 18, 1990

The breast implant issue is moving quickly, fueled by biased and inaccurate media coverage. . . . They have focused on a small number of anecdotal, sensational stories, not on science and reason. . . . The time to act is now . . . the well-being of your patients is your primary concern. *You are their advocate.*

—letter from the American Society of Plastic and Reconstructive Surgeons leadership to members, fall 1991

I want to stress that the FDA is not opposed to breast implants. We recognize the value of these devices, particularly for women who face a devastating and potentially disfiguring disease such as breast cancer. We *want* to see safe breast implants available for all women who need them. But the manufacturers of silicone gel implants have failed thus far to provide adequate evidence that they are marketing a safe product. . . . We owe it to the American public to see to it that these questions are thoroughly investigated.

—FDA commissioner David Kessler, January 6, 1992

> Not since the Dalkon Shield was taken off the market in 1974 have women been so frightened and confused by Federal officials raising safety concerns about a medical device. . . . Along with the fear . . . came anger that the manufacturers and plastic surgeons and the regulatory agency had, at worst, duped them, and at best bedeviled them with contradictory or inconclusive announcements.
>
> —*New York Times*, January 19, 1992

The "breast implant crisis," the "breast implant calamity," a "corporate tragedy," "silicone-gate," and the "breast implant circus"[1]—these appellations illustrate how much conflict has surrounded breast implants, the intense emotion the subject evokes, the range of individuals and institutions involved, and the various perspectives from which they view the issue. The word I use to describe the events recounted in this chapter is "controversy."

This term reflects my interest in what sociologist Joseph Gusfield calls the "drama of public action."[2] That is, in this chapter I explore the public struggle over implants, those points at which the major players came together to enact a sort of theater. In particular, I look at three such events: a congressional hearing held in 1990 and two FDA hearings, one of which took place late in 1991, the other early in 1992. Each hearing was the setting for the performance of certain set pieces. The plots of these dramas revolved around the fight to define the meaning of the breast implant problem and thus to control the outcome of the controversy. In scripts written by the interests and values of the players, the hearings became staged disputes over the information that could be counted as evidence, and then, a series of morality plays over its interpretation.

ACT ONE: THE WEISS HEARING (AND SOME PROLOGUE)

On December 18, 1990, Congressman Ted Weiss, Democrat from New York and chairman of the House Subcommittee charged with oversight of the FDA, held a hearing that posed the question "Is the FDA Protecting Patients from the Dangers of Silicone Breast Implants?" Weiss's introduction set the tone for the testimony to follow. He reviewed the current situation with implants, stating that two million women in the United States had undergone implant surgery and that, contrary to the public's belief, the FDA had never approved the devices as safe and effective. He accused the FDA of deliberate delay in addressing the issue of implants, warning that as the agency's inaction dragged on, "implant surgery continues on hundreds of thousands of women, many of whom are not warned about the potential dangers."[3] Clearly, while the hearing had as its stated purpose a review of the FDA's actions and procedures, it was predicated on an assumption of harm. Weiss spoke with assurance about the number of American women who had received implants, the lack of informed consent for the procedures, and the "potential dangers" of implants, but all were issues still to be contested.

That the FDA had never approved the devices was a fact, however. Although silicone gel breast implants had been available since the early 1960s, it was not until 1976 that Congress, with passage of the Medical Device Amendments to the Food, Drug, and Cosmetic Act, granted the FDA the authority to regulate them. The 1976 law required that the FDA place all medical devices into one of three regulatory classes. While placement in the first two classes required manufacturers to meet only minimal standards—mainly that the devices were not mislabeled or adulterated—manufacturers had to demonstrate that devices designated as Class III were "safe and effective" before they could be marketed.[4] All medical devices, new and existing, were to be assessed by advisory panels of experts appointed by the FDA. The panels would review information about the devices provided by the manufacturers, listen to the comments of other interested parties, and make classification recommendations to the FDA. Later, similar panels would hear scientific evidence about Class III devices to determine if the devices were safe and effective.

By 1990, silicone gel breast implants had been reviewed by several such advisory panels, composed largely of plastic surgeons. At each of these reviews—two in 1978 and one in 1982—the panel had recommended that breast implants be placed in Class II. Each time, however, the FDA had overruled the panel's recommendations, citing a paucity of valid scientific data with which to assess risk and benefit and indicating its intention to classify implants as Class III. In 1982 the FDA announced that implants would be put in Class III, but took no further action. Finally, in 1988, the agency made the final classification. Further delays—allowing time for public comment—were written into the law. Thus, it was not until 1990 that the FDA gave notice to the implant manufacturers that they would soon have to come before the agency to provide proof of the safety and effectiveness of the devices.

In his hearing, Weiss heard testimony from several groups of witnesses: women with implants, scientists and physicians involved in the issue, and FDA staffers. Finally, a representative of Dow Corning (the country's largest manufacturer of implants and supplier of implant raw materials) presented the company's position on implants and the current status of regulation.

The picture of implants that emerged during this hearing was almost entirely negative. Two of the three women with implants described, in excruciating detail, the "horror stories" of their experiences. They claimed that they had never been informed of the possibility of adverse effects. The third, however, representing a support group for breast cancer patients, credited implants with having "eased away much of the trauma of breast cancer."[5] The scientists and physicians characterized implants as flawed in design and manufacture, inadequately tested, and dangerous to health. The Dow Corning representative defended the device, noting his company's "great confidence that our breast implants . . . are safe and effective."[6]

The witnesses who identified themselves as victims of silicone implants used personal narratives of suffering, including the presentation of medical records and detailed chronologies of symptoms, to frame the problem of implants as one of multiple injuries. They focused on a purported lack of information, claiming that the facts about implants had been deliberately hidden. These facts were the clear, incontrovertible evidence for the dangers of implants, as evidenced by their own suffering. Like the victim witnesses, the professionals who testified against implants described them as inherently dangerous and suggested that the problem was one of legal malfeasance.

The defenders of implants, not surprisingly, described different problems. The woman who had had implants for reconstruction used personal narrative to tell a story of restoration. She argued that the risks of implants should be weighed against the benefits for thousands of satisfied women. Her testimony—through the claim that objections to implants were based on a belief that they were used by vain or selfish individuals—constructed implants as a problem of misunderstanding and prejudice. Because she emphasized the reconstructive uses of implants, her testimony linked implants and breast cancer, thus positioning the misunderstanding of implants as part of the problem of breast cancer.

The Dow Corning representative argued that implants had been proved safe and effective by thirty years of "solid science" (most of it performed by Dow Corning and protected as proprietary information) and by the one million satisfied women using Dow Corning implants. He framed the problem of implants as one of persecution by an overzealous government agency. Rejecting all claims of risk or uncertainty, he proclaimed Dow Corning's assertion of implant safety.

In questioning FDA staffers, Weiss focused on several issues: disparities between the opinions of individual FDA staffers and eventual agency policy; the history of delay in classifying implants; and the agency's failure to act on recommendations made by a recently convened expert panel. To understand why Weiss chose to highlight these areas, it is necessary to know something of what had already transpired.

In the spring of 1988, the FDA came into possession of a study performed by Dow Corning. The study showed that rats injected with silicone gel had developed a cancer called fibrosarcoma. Recognizing the importance of this information, the FDA had requested reviews of the research by its own staff scientists and scientists working in several other government agencies. This scientific review was inconclusive. While some experts opined that the cancer was likely to have been induced by a mechanism that did not occur in humans, others cautioned that the study should be taken seriously. (In general, the FDA policy on carcinogenesis is conservative: substances known to cause cancer in animals are to be restricted for human use.) In August of 1988 a high-ranking FDA scientist wrote a memo that concluded, "While there is no direct proof that silicone causes cancers in humans, there is considerable evidence to suspect that it can do so."[7] The FDA chose to make an official statement that reflected both scientific indeterminacy and political hedging: "One, although it is unlikely that the Dow Corning results would apply to humans, this possibility cannot be dismissed; and, two, if a cancer risk does exist for women with silicone gel-filled breast implants, it would be statistically small."[8]

The Dow Corning rat study became a matter of public concern in 1989 after its contents, and the FDA's reaction to them, were leaked to the press by the Public Citizen Health Research Group, a watchdog organization founded by Ralph Nader and headed by physician Sidney Wolfe.[9] At that point, Public Citizen had already petitioned the FDA to remove implants from the market, citing a variety of health risks. As the implant controversy intensified over the next several years, Public Citizen continued to play a highly public, and publicized, role.

Just weeks after the revelation of the rat study, the FDA convened another advisory panel. Although the panel accepted the FDA's conclusion that the risk of cancer from silicone implants was small and found no indication that it should necessitate the removal of the devices from the market, other evidence introduced at the hearing was more damaging. Several witnesses, including a lawyer who had recently settled a product liability case against Dow Corning, told the panel that the manufacturer was concealing from the FDA a large cache of documents relevant to the safety and effectiveness of the devices. On the basis of these allegations, the panel made several recommendations: First, the panel would reconvene in two months to consider any new data and to make recommendations about the specific information the agency should require from manufacturers seeking FDA approval of their devices. Second, with the cooperation of the FDA, the manufacturers, and the plastic surgeons, a national registry of breast implant recipients would be established. Third, the agency would work with the manufacturers, plastic surgeons, and consumer groups to develop a mandatory informed consent and patient education program. Finally, the panel directed the agency to keep the panel members, the public, and all other interested parties informed of new information as it became available.[10]

As testimony during the Weiss hearing revealed, however, several of these recommendations were left to languish. The chairman of the 1988 advisory panel, Norman Anderson, told Weiss of his frustration at the intransigence of the manufacturers, plastic surgeons, and the FDA. Within a month after the panel meeting, Anderson reported, he had written to the FDA to complain that the manufacturers seemed to be reneging on their commitment to develop a registry. As he told Weiss, the FDA had been equally as difficult, refusing to meet its commitments to the panel.[11]

The FDA representatives tried to answer each of the issues: the individual opinions of scientists were taken into account in formulating the consensus statement; the agency knew of some of the Dow Corning documents but had not seen all of them; a registry would be prohibitively expensive; the patient education materials were stalled over a dispute between two of the contributing parties; the agency was planning a scientific conference on the issue of silicone. These responses pointed to the conflicting demands placed on the agency. On the one hand, it was asked to provide scientific expertise, expertise always couched in the language of statistical probability. On the other, it was asked to be the final arbiter of policy, to make definite and authoritative statements involving risk. At the same time, the agency was a huge bureaucracy, subject to the demands of ever more products, yet provided with only limited resources to do its work.

The heart of the arguments embedded in the design and execution of the Weiss hearing lay in two of the congressman's statements: "Some scientists believe that 2 million American women have participated in a 20–year experiment on the safety of breast implants without knowing it";[12] and while "it was FDA that first highlighted this problem . . . my great frustration . . . is that, having highlighted and uncovered it, you have allowed yourself to be dissuaded from being more forthcoming in research and in undertaking to demonstrate safety and effectiveness on the part of the manufacturers."[13] (In statements to the press, Weiss provided more explicit political context to this accusation. He told the *New York Times* that "as was typical during the anti-regulatory climate of the Reagan administration, these efforts were blocked by higher-level officials.")[14]

Weiss's statements framed two problems: the danger of implants and the failure of the (Republican-controlled) FDA to protect health and safety. He looked to manufacturers for having caused the danger problem and to the FDA for having failed to solve it. The problem of the FDA's failure belonged to Congress. Direct responsibility for the delays lay with the agency but, more broadly, was an effect of the trend toward deregulation by conservatives in the executive branch. Weiss's implication was clear: it was only through congressional oversight (by the Democratic majority) that the FDA would act on the implant issue.

The Weiss hearing revealed many of the plotlines that would be played out as the public drama continued. The first was a series of queries about the safety of the devices: Did implants cause disease? What kind of evidence could be used to answer the question? Witnesses at the hearing presented various types of evidence—personal narrative, "solid science," history. The FDA had demonstrated a preference for scientific data as the standard for its decisions. It was clear, however, that the scientific validity of these evidences would be contested. Was one person's story of suffering data or anecdote? Could the manufacturers' claims for their proprietary data be trusted, or was science necessarily something public? What value did history have in making determinations of risk and benefit? The interests of the players involved in these questions were also revealed. For the opponents of implants, vindication and the possibility of legal recompense depended on broadening the definition of science to encompass personal narrative rendered publicly. The manufacturers and plastic surgeons, of course, opposed extending validity to stories of harm but approved it for demonstrating benefit among those women who were satisfied with their implants. Opponents and proponents clashed over the meaning of history. Advocates for implants, who had largely controlled that history, read it as positive evidence of safety; victims cited perfidy and distortion in rejecting its validity. The FDA, which would be called upon to make a final determination about the fate of implants, faced the public problem of which evidence to legitimize as scientific. Privately, it faced an ideological challenge to its authority.

The Weiss hearing stands out as a turning point in the controversy. While the opponents of implants were to hail it as the prod that finally forced action from the FDA, implant proponents were to criticize it as one-sided, staged for purposes of Weiss's self-promotion and anti-deregulation agenda. Although the proponents of implants were later to blame Weiss and this hearing for "politicizing" the issue of implant safety—by which they meant privileging media and consumer claims over expert and scientific claims—it is equally likely that the congressman was simply responding to the current zeitgeist. Implants had already become a public subject, with articles questioning their safety appearing in medical journals and the popular media and women claiming to have been harmed by the devices going to the courts to settle their grievances.

OFFSTAGE: PUBLIC ATTENTION TO IMPLANTS

Journals devoted to plastic surgery had, since the introduction of silicone breast implants, paid a great deal of attention to certain complications. By far the most heavily researched of these complications was capsular contracture, a localized condition in which the breast becomes hard to the touch and spherical in shape as the implant is surrounded by scar tissue.[15] Plastic surgeons wrote about the possible etiology of this complication, as well as how to treat it. They considered it an annoyance, but not dangerous. Other journal articles pointed to the possibility of serious systemic complications linked to silicone. Liquid silicone injections were known to cause silicone granulomas (benign tumors) and to lead to the destruction of local tissue.[16] Long-term reports from Japan alleged the development of immune disease following such injections.[17] In the early 1980s, case reports that linked silicone breast implants and autoimmune disease were published.[18] Around the same time, research suggested that silicone products underwent degradative change in the body, that the material was not as inert as had been thought.[19]

Popular journalism focused on the negative effects of silicone implants in the lives of real women. For example, in June of 1988 *Ms.* magazine published "Restoration Drama," in which author Sybil Niden Goldrich recounted her own experience with implants.[20] Goldrich asserted that although she had carefully researched the product before consenting to the procedure, "nothing in my research suggested that this 'simple' procedure would turn into five operations, over a period of ten months, requiring more than fifteen hours under anesthesia and countless days of pain and discomfort."[21]

Goldrich focused her critique on many of the same issues that later were brought to prominence in the Weiss hearing. She emphasized how little information about implants was available to women, implying that there was deliberate attempt by surgeons to keep their patients uninformed. She also criticized the FDA for its system of collecting and disseminating reports of device failure. The central problem, she argued, was one of definition. Manufacturers and surgeons preferred a limited definition of "failure": "a 'broken' implant—one literally damaged in the manufacturing." For women, the recipients of the implants, "failure" was a broader and more common phenomenon: "A failed implant is one that causes her pain or discomfort or requires its removal or change."[22] The distinction challenged both the assumption that it would be up to the experts—physicians and regulators—to control the meaning of failure and the manufacturers' assertion that plastic surgeons were their customers, while women were simply the "wearers" of the devices.[23]

After the article appeared, Goldrich was contacted by a number of women who had had similar experiences with implants. Along with one of these women, Kathleen Anneken, Goldrich founded one of the first advocacy-support-information groups for women with implants. Goldrich and Anneken named their group Command Trust Network (CTN). The name reflected the group's emphasis on the need for patients to demand full and clear information from their physicians—and their belief that trust would come only with such full disclosure. By 1990, CTN had achieved enough prominence that Goldrich was one of the women asked to testify at the Weiss hearing. As the events of the breast implant controversy unfolded, CTN continued to play an important role.

In December of 1990, just before the Weiss hearing, CBS television aired a segment about breast implants on *Face to Face,* a weekly magazine show hosted by Connie Chung. The story introduced several women who believed that they had been harmed by implants. It suggested that there was a "pattern of illness" among implant recipients and featured a statement by a doctor who implicated silicone in diseases of the immune system.[24] For many women with implants the broadcast was a consciousness-raising event, the first time they were offered a plausible explanation for the symptoms plaguing their lives. Those who became active in the movement against implants proclaimed it as an example of truth and journalistic integrity. To the manufacturers and plastic surgeons, however, the show represented the height of irresponsible, one-sided coverage and hysteria-generating sensationalism.[25]

The third site of public attention to implants was the courtroom. Recipients of silicone breast implants had been suing implant manufacturers since the 1970s. These suits, most of which were settled out of court, alleged product liability.[26] (Product liability claims relate to harm resulting from the failure of a product to function as intended under conditions of normal use.) In the most publicized of these suits, a 1984 case, Maria Stern sued Dow Corning on the grounds of product liability and fraud, claiming that her autoimmune disease had been caused by implants that had ruptured and leaked silicone into her body. In the course of pretrial discovery, the Stern legal team traveled to Dow Corning headquarters, where they were given access to internal documents. On the basis of these documents, and expert testimony linking silicone to systemic physiological harm, Stern won her case and was awarded $211,000 in actual damages and—more ominously for the manufacturer—$1.5 million in punitive damages. Later, during a lengthy appeal process, Dow Corning offered Stern a cash settlement with the proviso that all court papers (including the internal documents) be sealed and all experts remain silent as to the content of their testimony. Stern accepted this offer.[27] These documents were the ones that later raised the suspicions of the FDA's advisory panel when their existence was revealed by one of Stern's attorneys during testimony in 1988.

These constructions of silicone breast implants represented them as dangerous, yet simultaneously other public sites painted a very different picture. My review of women's magazines during the 1970s and 1980s shows that implants were presented favorably, as one of many consumer options available to modern women to enhance their beauty and improve their lives.[28] As in the controversy, these articles cite a variety of kinds of evidence, relying most heavily on personal narratives of transformation and interviews with authoritative and sympathetic plastic surgeons. Once the controversy broke, however, circa 1990, the women's magazines became some of the most vociferous purveyors of silicone horror stories, eliding their former celebration of "choice" with accusations of (male) establishment perfidy.

THE FDA ACTS AND OTHERS RESPOND

On April 10, 1991, the FDA published a final notice requiring implant manufacturers to submit premarket approval applications demonstrating safety and effectiveness. According to statute, the manufacturers now had ninety days to file these applications. After a scientific and bureacratic review of the applications by the FDA staff, the FDA would convene yet another expert advisory panel (scheduled for November of 1991), and then, based upon the panel's recommendations, the FDA commissioner would issue a final decision by January 6, 1992.

Spurred by the announcement, the American Society of Plastic and Reconstructive Surgeons (ASPRS) began a campaign to ensure the continued availability of the devices. As the largest professional organization of plastic surgeons in the United States, the ASPRS had taken an active but discrete role in the breast implant issue. Now, however, the society acted to greatly increase its public visibility. In the fall of 1991, the ASPRS levied a "mandatory special assessment" of just over $1,000 from each of its members. The money would be used, ASPRS president Norman Cole explained, to "enable the Society to respond quickly to this crisis."[29] The crisis lay in the "biased and inaccurate media coverage" of the breast implant issue and the influence of implant opponents on the "political and regulatory climate."[30] By raising questions about the safety of implants, the opponents were threatening the well-being of patients and the autonomy of plastic surgeons. The assessment was necessary to counter the current climate, "to help ASPRS to ensure that . . . accurate information is available to consumers and that all points of view are aired."[31] The ASPRS formed a "Breast Implant Crisis Task Force" to coordinate the society's response. The strategy of the task force was to move on several fronts: in the media and in the legislature to gain public support, and, privately, through the application of political influence.

As Cole had indicated, the organization's primary aim was to communicate information favorable to its position. Thus the first part of the strategy was to devise a comprehensive media effort. This effort included both internal publications and messages directed at the membership, and external publications, including a newsletter sent to patients. In addition, the ASPRS worked to enlist the popular media, suggesting that individual surgeons make themselves and their "especially articulate patient[s] who are willing to 'go public'" available for interviews. (The society supplied these practitioners and patients with media guides and prepared lists of "Key Message Points.")[32] At the end of October, the ASPRS ran an "advocacy ad" in major newspapers serving "key congressional districts." The society characterized the advertisement (described in the introduction and discussed in chapter 6) as "us[ing] the experiences of four women with breast implants to illustrate the benefits of the devices and the consequences of removing them from the market or restricting their use."[33]

As part of their legislative strategy, the society urged their membership to write to elected representatives in Congress. Several ASPRS mailings included form letters that surgeons could replicate. The society suggested that surgeons contact their implant patients and ask them to write letters as well. ASPRS-produced newsletters directed at patients also emphasized the letter-writing strategy, and included "tips" for the contents of such letters.[34] In addition, the task force wrote directly to the ASPRS membership to solicit contributions (minimum suggested amount: $500) for "PlastyPAC," the organization's national political action committee. The PAC was necessary, the task force wrote, "because we must be able to show our appreciation in a tangible way to congressmen who are receptive to our message."[35]

The media and legislative strategies converged in October of 1991, when the ASPRS sponsored a "Washington, D.C. Fly-in" for surgeons and patients. During the event, which was heavily promoted in ASPRS press releases, the society made arrangements for implant proponents to meet with legislators and "urge [them] to preserve a woman's right to make an informed choice about breast implants."[36]

The ASPRS strategy of attracting public attention was supplemented by a quieter effort to mobilize private political support among Washington influence brokers. The ASPRS hired lobbyists with links to both the FDA and the White House to plead its case.[37] These lobbyists used their connections to access with men as highly placed as John Sununu, then chief of staff to President George Bush.[38] Their primary message was that implants must remain available. A secondary message was directed against the commissionership of David Kessler. Reportedly, the ASPRS demonstrated its understanding of the importance of "private" information as evidence by passing a rumor that Kessler was biased against implants because his wife had had health problems because of the devices. Kessler emphatically denied this rumor.[39]

Use of public relations strategists and influence brokers was not limited to the plastic surgeons. The manufacturers hired their own advocates. On the other side, both Public Citizen and the Association of Trial Lawyers of America worked hard to sway decision makers. Like the surgeons, implant opponent groups organized events that attracted press attention in order to promote their anti-implant position.[40]

When the FDA deadline arrived in July of 1991, the agency had received applications from seven implant manufacturers. The agency rejected out of hand the applications from three manufacturers, citing a lack of "information based on human studies," but emphasized that its decision was not to be interpreted as an endorsement of the claim that implants were dangerous. Rather, "it means that the information provided by the manufacturers is not sufficient at this time to permit the agency to evaluate the safety and effectiveness of the devices."[41]

The FDA accepted for review applications for seven products (separate applications were required for each model of implant) from four manufacturers: Dow Corning, McGhan Medical, Bioplasty, and Mentor. The agency's move to proceed with the review was, as with the rat study several years earlier, not without internal dissension. Several FDA staffers recommended that all of the applications be rejected due to "major methodological flaws," or problems of evidence, in the studies they reported. For example, critics cited the short follow-up times, the large numbers of subjects lost to follow-up, a lack of studies looking at systemic effects, and several implant models for which there was no human research.[42] Reportedly, the agency's decision to proceed with the seemingly inadequate applications was a reflection not of scientific judgment but of political concern. According to a memo from congressional staffer Diana Zuckerman to her boss, Congressman Weiss, the FDA had accepted the applications because it feared that rejection would result in "a lengthy appeals process instigated by the manufacturers." In addition, the agency recognized that without the regulatory process, information would be lost: "The companies would not continue their studies of implant safety if the [applications] were not filed, thus limiting information that could eventually be made available to the public."[43] The agency's acceptance of the applications ensured that part of the battle, at least, would take place in the open.

ACT TWO: FDA HEARING, NOVEMBER 1991

The meeting of the FDA's expert advisory panel was held over three days in November of 1991. The atmosphere was highly charged. Both sides had worked hard to promote their cases to the public and in the corridors of power. The media had deluged the public with stories about implants; some were well reasoned and factual, others sensationalistic. Advocates for implants and their opponents had been mobilized.

The composition of the November panel reflected the FDA's awareness of the need to balance political and scientific concerns. The panel was composed of two groups: nonvoting consultants or representatives and voting members. The former included several plastic surgeons—a distinct contrast from earlier panels, when plastic surgeons had predominated in the ranks of voting members—two well-known implant opponents who had testified as plaintiff experts in a number of product liability cases, a representative from industry, and several consumer representatives. Voting members included physicians trained in several specialties (pathology, radiology, oncology, orthopedics, internal medicine), a psychologist, a lawyer-ethicist, an epidemiologist, and a medical sociologist.

The FDA's charge to the panel was tripartite: first, the meeting was to provide a public forum for the airing of a range of views;

second, the panel was to review the manufacturers' applications and evaluate "whether the scientific data on [implants] . . . are sufficient to support the approval of the applications"; and finally, if the applications were judged inadequate, "the Panel will be asked to advise FDA on whether or not the continued availability of the device is necessary for the public health."[44]

In his opening address, FDA commissioner David Kessler asked for the panel's guidance on the question of the public health value of implants: "Would it be desirable, from a public health standpoint, to ensure at least limited access to the implants if their safety and effectiveness data were inadequate?" He alluded to the possibility of different policy outcomes for reconstruction and augmentation: "If continuing access were desirable, would this be true for all patients? Or would certain classes of patients suffer particular hardship if these products were not available?"[45] Kessler also took the occasion to assert the FDA's claims of authority over implants and to indicate which kinds of evidence would be accepted by the agency.

On the first day the panel heard testimony from a variety of professional and consumer groups, as well as individual medical practitioners and patients. Professional societies including the ASPRS, the American College of Surgeons, the American Medical Association, and the American College of Radiology were joined by breast cancer support groups, the American Cancer Society, and the Susan G. Komen Foundation in supporting implants. Groups like CTN, American Silicone Implant Survivors (AS-IS), Coalition of Silicone Survivors, the Boston Women's Health Book Collective, and Public Citizen opposed them.

In their statements, the implant advocates emphasized the value of the devices in helping women to recover after mastectomy. As one woman said, "Now when I look into the mirror, I do not have to think cancer patient anymore. I feel whole again."[46] Reconstruction was not a matter of "vanity," they argued, but of ameliorating "shame and fear of rejection."[47] They portrayed implants as an issue of personal freedom: "Why are we here today? One of the answers is simply freedom of choice. A person's freedom to swing his arm ends where another person's nose begins. A women's breast implant will not affect another individual. The choice of an implant stops far short of another person's nose."[48] Implicitly, the advocates linked choice and personal responsibility for that choice. As one woman noted, "I never had the unrealistic expectation that a foreign object placed in my body was going to function better than my body does."[49]

Advocates spoke of continued availability as a matter of fairness: "The misfortune of these few should not deny millions of other women an opportunity to choose."[50] Others argued that restricting implants would have dire consequences: women with implants would be discriminated against by insurance companies, mastectomy patients would suffer emotional distress, biomedical research would be stymied, and more women would die from breast cancer.

Although the advocates focused on reconstruction, they also presented arguments against making a policy distinction between reconstruction and augmentation. Perhaps the most powerful argument was that such a distinction would be emotionally devastating to reconstruction patients: "Please do not insult us by deciding implants are acceptable for women with breast cancer, but not for our sisters who choose augmentation. Are our lives not as valuable? Do you expect that we will not live?"[51]

Even without their verbal statements, the implant advocates made a strong argument. The argument was implicit in their numbers, in the vast proportion of those who had had implants for reconstruction, in the explicit linkage of implants and breast cancer, and in the participation of influential women like the wife of Florida senator Connie Mack. The messages they conveyed were that implants were most commonly used for reconstruction, that they were vital for that purpose, and that articulate, well-educated women would fight for access to the devices.

The testimony of witnesses from professional societies shored up the testimony of individual advocates. These groups emphasized the value of implants in cancer rehabilitation and contended, without specific evidence, that a ban would dissuade women from seeking early detection of and treatment for breast cancer. These professionals portrayed the choice of implantation as a matter to be decided by a woman in consultation with her doctor.

The proponents of continued availability framed the problem

of implants as a problem of breast cancer. This framing was both implicit (in the identification of witnesses as breast cancer survivors) and explicit (for example, one woman commented that money being spent to regulate implants should go to breast cancer research instead).[52] With the problem defined this way, implants became entangled with the politics of breast cancer, a politics of an increasingly powerful Washington lobby that was demanding more attention to breast cancer and more government funding for the attention.[53]

Self-identified "silicone victims" and other opponents of implants, however, made very different arguments. The "right to choose" was not absolute: a woman's right was "to choose a safe, tested, and approved product, not a poorly designed experimental product."[54] Like the implant advocates, opponents spoke of freedom of choice and personal responsibility. In their eyes, however, "freedom of choice was a possibility that manufacturers and plastic surgeons denied patients for the many years they withheld information about the down side risk of breast implants."[55] They questioned whether informed consent was possible even now because, they claimed, the unknowns about implants were still too great: "Informed consent is only as good as the information available to those making the decisions."[56]

The silicone victims also responded to the claims of plastic surgeons and implant advocates that they, the victims, were a small minority. There may be only a few of us here today, said several women, but we are the "tip of the iceberg."[57] As supporting evidence, representatives of victim/survivor support groups noted their absent constituencies, claiming that they were representing many other women who could not travel to the hearing because they were too ill or because their illnesses had bankrupted them. CTN founder Sybil Goldrich, for example, reported that her group had eight thousand members in the United States.

The rhetorical strategy of the victims was to rely on their personal narratives, on the presentation of photographs of their mutilated bodies, and on phrases like "silicone time bomb" and "guinea pig" to frame the meaning of implants as suffering and perfidy. In contrast to the proponents of implants, who used percentages when speaking about complications, opponents like Sidney Wolfe of

Public Citizen applied those percentages to the estimated total number of women with implants, noting that even a small rate when multiplied by a large number meant thousands of women in pain and distress. The victims emphasized their betrayal by the plastic surgeons and manufacturers. "If I were told the truth about this product lasting only a few years, none of us would have agreed to this barbaric way of treatment," said one woman, her very grammar eliding the difference between her experience as an individual and the suffering of the group. "Try to understand how we trusted and believed, but we were definitely misguided for sake of money."[58]

On the second day and into the third, the panel reviewed the applications submitted by the four manufacturers. In introducing this scientific review, the FDA urged the panel to consider the following issues: the physical and mechanical characteristics of implant materials; the chemical composition of the materials; the metabolic fate of silicone in the body; the potential for silicone to have biological toxicity; the effects of silicone exposure on the immune system; the incidence of adverse effects in women with implants; the extent to which implants interfered with mammography; and the measurement of functional and psychological benefit. The FDA's own position was clear: the applications were inadequate.

The manufacturers presented material meant to address each of these issues in the scientific language favored by the FDA. As questioning by the panel revealed, however, there were many knowledge gaps in their presentations. The physical and mechanical testing that had been completed did not seem to replicate the in vivo conditions of implant use. The exact chemical composition of the implant materials was unclear—vagueness due either to a lack of analysis or to the manufacturers' reluctance to make their trade secrets public. The clinical studies were vulnerable to criticisms of bias and flawed design. Studies of benefit were not rigorous, lacking control groups or long-term follow-up. Finally, the manufacturers were unable to answer some basic questions, such as the total number of women with implants in their bodies.

By nearly unanimous vote (the only opposition came from the radiologist on the panel), the panel voted to reject all of the manufacturers' applications on the basis of insufficient information to prove safety and effectiveness.

In the next phase, as Commissioner Kessler had requested, the panel set out to determine if implants should remain available while manufacturers worked to gather the data needed to provide proof of safety and effectiveness. In doing so, the panel considered whether there was a public health need for the devices that would outweigh the possible (though unproven) risks. An FDA staffer offered some guidance to the panel members, urging them to judge "first, whether the patients would suffer undue hardship if these products were not available. Second, whether the patients would be exposed to an unacceptable risk if we did allow continued access while additional data are being collected; and third, whether we can reasonably distinguish between patient groups."[59]

The panel members debated these issues. A consumer representative noted that, strictly speaking, all implantation was medically unnecessary.[60] The psychologist argued that the importance of breasts exemplified an "unconscious ideology . . . [that] equates femininity with beauty."[61] The representative of a breast cancer support group argued that there was a great public need for the devices, and that removing them from the market would have a ripple effect of harm.[62] The radiologist who had voted to approve the applications asserted that if women could not have access to implants for reconstruction, their compliance with mammographic screening recommendations would decrease.[63] The plastic surgeons echoed the belief that there was a public health need for the devices. Several panel members noted that the suggestion of different policies for augmentation and reconstruction patients seemed to pit women against one another. Two female panelists agreed that they were uncomfortable with establishing a "hierarchy of need" between reconstruction and augmentation.[64] Another panel member drew a distinction between a finding of harm—which the panel had not made—and a finding of inadequate information—which it had.[65] Although implants seemed to pose no immediate threat, the panel agreed, neither was there full information about the possible effects of long-term exposure.

In an attempt to balance these views and arguments, the panel made the following recommendations: The review process would be extended. Implants would remain available to all patients, but only under certain conditions, including the continuation of concurrent preclinical, clinical, and epidemiological research, the institution of a mandatory informed consent procedure, and the establishment of a registry for patients not taking part in clinical trials. The FDA was to be responsible for guiding and monitoring the research and manufacturers' adherence to the recommendations. Any violations would cause the FDA to institute severe restrictions on the availability of the devices.

The panel's deliberations and recommendations reflected competition among the different ways that implants had been framed in testimony. While certain nonvoting consultants articulated arguments that were identical to those expressed by partisan witnesses, the voting members of the panel expressed more ambivalence. Their final recommendations showed an acceptance of the FDA's claims of uncertainty—militating against acceptance of the applications—but also an acceptance of the implant advocates' construction of the problem as one of breast cancer and the argument of plastic surgeons and other advocates that to make a policy distinction between reconstruction and augmentation was morally unacceptable. The panel's mandate to focus on the sufficiency of information in the applications allowed it to sidestep any consideration of the broader issue of implant safety. In so doing, it avoided having to endorse any one group's claims.

In a statement made the day after the panel issued its recommendations, Commissioner Kessler "pledged that FDA would require manufacturers to provide the information needed to answer questions about these products, and that the agency would take into account the needs of women who now have the implants or who might desire them in the future." However, the statement cautioned, "the committee's recommendations are not binding upon the FDA."[66]

Public reaction from the manufacturers and the ASPRS was laudatory. A Dow Corning executive said that the company was "truly gratified that this panel recognized the public health need" for the devices.[67] A statement from the ASPRS called it a "sensible, compassionate decision,"[68] and the society soon announced an allocation of $500,000 for breast implant research.[69]

In private, however, implant proponents recognized that the

situation was still critical. "The advisory panel's mixed vote gives the FDA wide latitude in its decision," Dow Corning CEO Daniel Hayes warned in a letter to plastic surgeons. "The FDA still has the option of restricting the availability of the device. . . . Commissioner Kessler's comments after the hearing made it clear that *he* was the one who would make the final decision."[70] Both Hayes and the ASPRS urged plastics surgeons to continue the fight for implants.

What Kessler would have decided will never be known, however. Weeks after the November FDA hearing another institution proved willing to make a determination about the safety of implants. That institution was the court, which showed itself to be receptive (as indeed it had been in the past) to the implant opponents' construction of the problem of implants.

THE PLOT THICKENS: ENTER THE DOW CORNING DOCUMENTS

On December 13, 1991, a California jury found in favor of Mariann Hopkins, the plaintiff in a product liability case against Dow Corning. In her suit, Hopkins asserted that silicone that had leaked from two pairs of defective implants manufactured by Dow Corning had caused her to develop connective tissue disease, and that the company had committed fraud by covering up its knowledge of defects in the product and the likelihood of harm resulting from those defects. The jury awarded Hopkins $840,000 in compensatory damages and $6.5 million in punitive damages.[71]

The evidentiary crux of the Hopkins case was a collection of Dow Corning documents relating to the implant models developed by the corporation in the mid-1970s. Although many of the same documents had been used in the Stern case in 1984, they had been under court seal since the settlement. Given the increased public awareness surrounding implants, the Hopkins case received heavy attention by the press. When some of the documents were displayed in open court, a reporter copied down many of the most incriminating points.[72] Eventually, copies of these documents made their way to Norman Anderson,[73] who had been chairman of the FDA advisory panel in the 1980s and had served as a voting member of the most recent panel. (Anderson had been openly critical of the implant manufacturers in his testimony at the Weiss hearing.)

Anderson wrote a strongly worded letter to FDA commissioner Kessler, alerting him to the existence of the Dow Corning documents and accusing the manufacturer of willful and deliberate obfuscation, withholding of data, and prevarication in its dealings with the FDA, the FDA advisory panels, and the public.[74] In a statement to the press, Anderson said, "I think it is time to stop trusting and take implants off the market."[75]

Revelation of the documents' existence increased the public storm. Dow Corning's Hayes sent an angry letter to Kessler, denouncing both Anderson and his allegations.[76] Weiss called for an investigation of Dow Corning by the Justice Department.[77] The FDA sent investigators to search Dow Corning's records and demanded a full accounting of all documents pertaining to breast implants.[78] In January 1992, Public Citizen again called for a complete ban on implants and accused the FDA of complicity with the manufacturers in misleading the members of its own advisory panel.[79]

At the center of the storm was a collection of approximately ninety documents, including both internal memoranda and scientific reports. Most of the material dated from the mid-1970s, when Dow Corning, under pressure from several new competitors, formed a "Mammary Task Force" charged with designing, producing, and marketing a new implant under a five-month deadline. The documents included internal memos from company employees expressing concern about quality control problems, instructions to the Dow Corning sales staff recommending that they rinse visible gel bleed off sample implants before showing them to prospective customers, and letters from several surgeons complaining of implants that had ruptured during implantation procedures. The new line of implants was not tested in animals for any length of time before being put on the market, but there were reports from several animal studies that showed silicone was not biologically inert. A later judicial review of these documents characterized them as demonstrating that Dow Corning had "rushed the development of silicone gel implants, failed to adequately test the implants, and ignored knowledge of adverse health consequences associated with the implants."[80]

Dow Corning contested the validity of the documents as evidence, arguing that the documents were "the airing of differences and interpretations. They involve individual points of view, but never were the consensus of the company."[81] The documents showed once again the power of private information made public. As the controversy continued, however, the circumstances of their revelation would make them vulnerable to challenges to their value as evidence.

The jury's decision in the Hopkins case validated the silicone victims' construction of the problem of implants as one of suffering and perfidy. It also bestowed legitimacy on the scientific hypotheses of the doctors and scientists who shared the victims' opposition to implants. By legitimizing the victims' version of the problem, the court ensured that the problem, and the relevant evidence, would have to be considered by the FDA.

INTERMISSION: THE MORATORIUM

On January 6, 1992 (the date by which the FDA was required to make a final decision about the manufacturers' applications), Commissioner Kessler announced a moratorium on the distribution of silicone gel implants and a request that surgeons stop all use during this period. In a statement explaining his decision, Kessler cited "new information about implants that amplifies our concerns about their safety." Although he did not mention the newly revealed court papers explicitly (the documents had not yet been made publicly available), he alluded to their existence.

Announcement of the moratorium increased the flood of reaction that had begun with the revelation of the Dow Corning documents. Several countries, including Canada, Germany, Spain, France, Austria, and Italy, followed the FDA's lead and imposed their own moratoria.[82] The ASPRS urged women not to panic and demanded that the FDA release the new information.[83] ASPRS president Cole said the decision was "very disturbing, and sends a very negative message to patients that there is something wrong with implants."[84] Dow Corning held a press conference to reiterate the company's belief in the safety of implants[85] and began to do damage control on some of the specific allegations.[86] An AMA spokesman predicted "absolute hysteria among women.[87]

Those who had been agitating for the FDA to take action also reacted. Weiss stated that there was "reason for concern, but not necessarily alarm,"[88] and urged that Dow Corning be investigated for violations including misbranding of its product and scientific misconduct.[89] Public Citizen accused some surgeons of ignoring the ban and continuing to use their own stock of implants.[90] CTN, which had been calling for a moratorium since December, applauded the decision and attributed it to the pressure brought to bear by its membership.

On January 20, the FDA ordered Dow Corning to make the documents public. Dow Corning resisted. The FDA threatened that if the company did not release the documents, the agency would do so on its own. A Dow Corning executive called the situation "a true travesty, a media circus instead of a true and impartial scientific review."[91] On January 21, Dow Corning agreed to release the scientific reports. CEO Hayes said he hoped that the FDA's request for release of the scientific evidence represented "a sign that the FDA's review of our product will return to the scientific arena, rather than rely on the innuendo, anecdotes and non-scientific internal correspondence that has typified the review process up to this point."[92] On January 23, Dow Corning agreed to release the internal memos as well. In releasing the documents, Dow Corning called for "a review of the science by a panel of experts qualified in the fields of rheumatology, immunology and inflammation, perhaps under the auspices of an independent government agency like the NIH."[93] On February 10, the company announced the appointment of a new CEO, Keith McKennon. The *New York Times* reported that Dow Corning was standing by the safety of implants but "has taken a new attitude on scientific questions and public opinion."[94]

In its report on the Dow Corning documents, released in February of 1992, an ASPRS committee charged by the society with reviewing the new information found no reason to "alter their assessment of the safety of these devices . . . [and] no information to materially change their view of implant risks and benefits." The authors of the report questioned the validity of the documents as evidence, noting that there was a "very great difficulty inherent in arriving at any scientific conclusions based upon review of information selected by attorneys to prove an allegation. *Science* is the gathering of ALL evidence before attempting a conclusion. To begin with an allegation then select only the evidence that will support the predetermined conclusion is not science."[95] Soon after releasing the report, the ASPRS sent a letter to its members urging that they demand Kessler's removal from the implant decision. "The oath of a doctor is to do no harm," read the letter, "but Dr. David Kessler . . . is doing immeasurable harm—to the public and to our patients . . . remove Commissioner Kessler because of his gross mishandling of this issue and his refusal to disclose the important medical evidence."[96]

Once the Dow Corning documents were made public, a flood of women with implants sought legal counsel.[97] On January 24, a class action suit was filed in Cincinnati against seven implant manufacturers. The suit alleged that "the manufacturers designed, manufactured, and distributed defective products; negligently carried out the manufacturing, testing and distribution of the implants; failed to adequately test the implants . . . and failed to warn silicone breast implant patients of that fact; and negligently and fraudulently represented to silicone breast implant patients, their spouses, the FDA and to medical providers that the products were safe."[98] A judge certified the class on February 14.

ACT THREE: FDA HEARING, FEBRUARY 1992

On February 18, the FDA reconvened its advisory panel so that the members could reconsider their recommendations to the FDA in light of the newly discovered information. In reaction to recent events, the composition of the panel had been changed: two rheumatologists, a pathologist, and a radiologist had been added as nonvoting consultants, and Norman Anderson had been demoted to nonvoting consultant.

In his charge to the panel, Kessler sketched the areas of concern that had prompted him to call for the moratorium. These included the Dow Corning documents, which "raise questions about the adequacy of quality control and product testing"; "information from clinicians about issues involving rupture, leakage, and bleed"; and "additional information including reports from rheumatologists, which have strengthened the possible connection between breast implants and inflammatory and autoimmune disorders."[99]

Kessler asserted that the FDA was not interpreting, or presenting, the Dow Corning documents as conclusive evidence, or even as scientific. Rather, he argued, the value of the documents was that they raised "sufficiently serious questions" about the thirty-year history of implants that manufacturers had claimed was proof of safety. In all, the new information reinforced the agency's construction of the problem of implants as one of uncertainty.

The manufacturers responded to the FDA presentation. Dow Corning, represented by its new CEO, Keith McKennon, displayed a conciliatory attitude, telling the panel, "I believe my and Dow Corning's overriding responsibility in this issue is to women who are using our implant devices." (His statement showed an important shift in the manufacturers' view of just who their customers were.) He stressed the company's desire to work with the FDA and the plastic surgeons to execute an integrated research plan. "We would like to move into a future that acquires all the data that are needed, and we are fully committed in every way to make that happen but we will need your help to do it."[100] Despite this attitude, Dow Corning continued to dispute the FDA's use and interpretation of the documents, arguing that the agency was trying to mislead the advisory panel. A company representative rebutted specific claims made by the FDA. For example, he explained that Dow Corning salesmen had been instructed to rinse off the implants before presentations not to disguise the existence of gel bleed but because bleed attracted dirt.[101]

Mentor, McGhan, and Bioplasty (the other manufacturers whose applications were pending) also made presentations. They argued that they were being unfairly tarnished by the revelations in the Dow Corning documents, that the contents of these documents should have no bearing on their own applications, that the FDA had misinterpreted the science, and that information about older implant models should not be used to make a decision about the implants to which the applications referred. Each manufacturer submitted data that it claimed proved that the risk of rupture and gel bleed was known and was much less than the FDA had suggested. Mentor claimed to have done extensive fatigue testing of the silicone rubber envelope. A plastic surgeon speaking as part of the Bioplasty team argued that the incidence of bleed and rupture had declined with innovations in implant technology: "It can be said of today's implant, these are not your mother's Oldsmobile."[102]

The FDA's claims for a possible association between implants and autoimmune disorders were disputed just as vigorously. All of the manufacturers argued that extensive animal testing had shown no results supporting such an association. The occurrence of autoimmune disease in women with implants was no higher than would be expected in the general population. Testifying as part of the presentation by Mentor, an immunologist summarized the mechanisms of silicone-linked immune activation that had been proposed in the literature and in testimony before the panel. He argued that none of these routes was supported by the currently available evidence. Only epidemiological evidence could provide the needed data, and, because of the rarity of autoimmune disease and certain problems of definition, such studies would be difficult to perform.

Plastic surgeons reinforced the attack on the FDA's presentation. Like the manufacturers, they disputed the agency's estimate of the rate of rupture. They conceded that a ruptured implant should be removed (and replaced), but argued that this was not because loose silicone posed any risk, but because removal would "alleviate . . . anxiety or fear the patient feels due to the speculation about the possible effects of implant rupture."[103] The surgeons downplayed the possibility of an association between implants and immune disorders, arguing that there was no supporting scientific evidence for such an association, only anecdote. "Nonscientists . . . find comfort in blaming something, regardless of how unsound that professed linkage may be," said one plastic surgeon.[104] Such was the case with the women who had been diagnosed with immune disorders after receiving implants.

The plastic surgeons' challenge to the FDA introduced another argument: that the suggestion of a link between implants and systemic health problems was a manifestation of "pathologic science"—science that came not from the clinic or the laboratory, but from courtroom testimony by paid experts, who distorted the truth in order to win large jury awards. The argument about pathologic science was a component of the claim that the breast implant issue had become entirely political. Manufacturers, plastic surgeons, and implant advocates argued that this politicization had begun in the courtroom, had been exacerbated by the Weiss hearing and by reports in the press (necessitating the response by the ASPRS), and now was reaching its apex in the kangaroo court of the panel meeting.

For the silicone victims and other implant opponents, however, the Dow Corning documents and the FDA's attention to the implant-immunity link were perceived as validation of long-held positions. Sensing victory, these groups shifted their focus, concentrating on convincing the panel that the new information about implants warranted a new, more restrictive recommendation. Sidney Wolfe of Public Citizen argued that the manufacturers and plastic surgeons could no longer be trusted to carry out the necessary research on implants, research they had several times promised but never delivered. CTN founder Sybil Niden Goldrich told the panel to "cut to the chase. . . . Now it is time for us to begin a transition to a world without silicone gel-filled implants."[105]

As the panel reached the deliberation stage, the FDA set the members two tasks: The first was to develop consensus statements that the FDA could use to answer consumer inquiries about silicone implants. The second was to make recommendations to the agency about its future policy toward implants—the "appropriate level of availability and the proper conditions to be imposed if the implants remain available."[106]

An FDA staffer spoke to the panel about the need for the agency to answer public inquiry. The FDA had received many questions from women who already had implants, she said. She asked the panel to consider a number of issues and reach a consensus based on the testimony that had been provided. The issues were implant lifetime, the correlation between adverse effects and the length of time the device had been implanted, capsular contracture, early detection of cancer, the use of mammography for detecting rupture, the proper course of action if an implant ruptured, and the fate and safety of silicone gel in the body.

The panel's discussion of these issues, which ranges over more than 125 pages of the hearing transcript, showed the difficulty of assessment given the paucity of data. The panel members quite quickly agreed, for example, that implants should not be considered lifetime devices, but they could not settle on an estimate of just how long the devices were likely to last. The issue of mammographic screening of asymptomatic women for rupture presented more dilemmas. If women were routinely screened, more rupture would be found. The panel members struggled to balance the unknown risk of living with escaped silicone in the body with the risks of increased exposure to radiation (in the mammogram) and the risks of increased surgery (if the ruptured implants were removed). Should the consensus statements alleviate unhealthy anxiety or raise healthy concern? How could they make an authoritative pronouncement when there was so much uncertainty?

The resulting statement showed the result of so much struggle. The panel urged women with implants to have regular checkups with their plastic surgeons or general physicians. It recommended that women have mammography for early detection of cancer when they reached the age suggested by current guidelines, but recommended against regular mammographic screening for rupture. If rupture were known to have occurred, however, the panel recommended that the implant be removed. The panel found that there was insufficient evidence to show a cause-and-effect relationship between implants and autoimmune disease, but recommended that women experiencing symptoms such as muscle and joint pain see their doctors. The panel also advised that further research was needed on the incidence of autoimmune disorders in the implant population.[107]

The panel's discussion illustrated how nuances of definition can shift meaning and just why the players in the breast implant controversy had worked so hard to promote their own ideas about the problem of implants. This panel of experts, immersed in the silicone gel breast implant issue for several days, and privy to all the studies and testimony, in the end produced nonconclusive, vague statements, and these only with great difficulty. (Perhaps the FDA set the panel this task as a form of subtle persuasion. The agency gambled that once faced with the uncertainties surrounding implants in its own decision making, the panel would be less likely to recommend a policy that left implants widely available.)

The policy recommendation stage of the proceeding was introduced by an FDA staffer who briefed the panel members on their options, which ranged from allowing the current situation of unrestricted implant availability to recommending a complete ban of the devices. He advised the panel to consider several factors: the "degree of therapeutic benefit . . . risk in relation to benefit . . . whether these benefits are so significant that despite the known and unknown risks these devices should still be available . . . availability and adequacy of alternatives."[108]

Once again, the panel's deliberations reflected the multiple meanings embedded in implants. There was broad agreement that a need for implants did exist. One minority saw implants as harmful. Another minority saw no risk. The plurality found uncertainty, and a disturbing lack of data. In the lack of data there was doubt, and in doubt, the possibility of danger. Debate thus focused on whether the need for implants outweighed the potential risk in all cases. Only one person, the nonvoting consumer representative, seemed to call for a complete ban on implants: "I don't think," she said, "we have to leave the device on the market because it's already there and it's popular."[109] Fairly quickly the panel voted to eliminate the option of a ban.

The choices were then limited to maintaining the recommendation made at the last panel meeting—widespread availability while studies were continued—or recommending some sort of restricted access. The plastic surgeons and the industry representative spoke in favor of maintaining the recommendation of the previous panel, arguing that the "new information" was irrelevant and should have no bearing on the panel's decision. For the voting members of the panel, however, the information presented had made a difference. One panel member noted that while the focus in November had been on benefit, the focus of this hearing had been on risk. "Because of the dimension of risks discussed at this meeting," she said, "I think that there is a need to limit those women who will be put at risk, and to ensure that the benefit they receive is justified by the risks that they court."[110]

The panel's final recommendation was the result of its focus on two issues: first, the recognition that the more restricted the access to implants, the fewer data could be gathered about the issues of safety still in question (and the less incentive there would be for the manufacturers to continue research); and second, the consideration of different risk-versus-benefit equations for reconstruction and augmentation patients.

If fewer women received implants, it would take the manufacturers longer to collect sufficient data. (And given the numbers needed to answer some of the epidemiological questions, it was questionable if such information could ever be gathered.) If no use were allowed outside of clinical trials, the manufacturers would be carrying the costs of the trials and the expenses of production without the profits garnered by widespread sales. Under such circumstances, members of the panel realized, manufacturers were likely to withdraw entirely from the implant market.

Although the minorities convinced of the extremes of safety and of danger still refused to make a policy distinction between reconstruction and augmentation, the plurality in the middle now proved ready to do so. The new willingness of the panel to make such a distinction was a reflection of the emphasis the FDA had placed on the potential dangers of implants and the panel's desire to "limit those women who will be put at risk," as well as the panel members' perception that the scientific evidence of benefit that existed for reconstruction was lacking for augmentation. The plastic surgeons on the panel attacked the distinction, calling it "judgmental, paternalistic . . . a ruling on the validity of cosmetic surgery."[111] Other panel members disputed this accusation: "I think the benefit to women who request augmentation is mitigated by certain risks that they experience that are not experienced by the breast cancer victim looking for reconstruction," said a psychologist arguing for a policy distinction. "Augmentation of the breast threatens its primary functions—its function of lactation, and its functions of sexual sensitivity and bonding with an infant."[112] Another panel member pointed to the lack of evidence of benefit in augmentation patients, concern about asymptomatic rupture, age differences between the two populations, and the risk of reproductive-related complications in younger women.[113] One of the harshest critics of implants, a nonvoting consultant who had served as an expert witness in a number of lawsuits against the manufacturers, recommended that in order to "minimize the duration of harm," implants should be allowed only for "older women . . . who have had breast cancer, who are in need of reconstruction, for whom there are no other alternatives."[114]

The panel's final recommendation, in effect, attempted to structure a compromise that balanced all of these concerns: Implants would remain available, but only through participation in FDA-approved clinical trials. For women seeking implants for reconstruction, there would be wide access to these trials. Augmentation studies, however, would be limited to a very small number of subjects.

The recommendation to restrict availability to clinical trials reflected the success of the FDA's framing of the problem of implants as one of uncertainty, and a willingness to accept the FDA's practice of placing uncertainty on the risk side of the risk-benefit equation. In making a policy distinction between reconstruction and augmentation, the panel went further, accepting some of the implant opponents' claims about danger. The language justifying the distinction represented a rejection of the argument made by implant proponents that such a distinction would represent a moral "judgment" of women's motivations.

Congressman Weiss applauded the panel's recommendation. Sidney Wolfe of Public Citizen criticized it, saying, "These devices were shown to be unsafe in both animals and humans. Now thousands of women will still be guinea pigs."[115] Reaction from those who had fought to maintain the availability of implants was critical but muted. The ASPRS argued that the panel's policy recommendation contradicted its consensus statement that there were no data showing a link between implants and autoimmune disease. In a press release the society said it was "disappointed that [the panel] chose to discriminate against the more than 100,000 individuals who seek the device for other reasons that are often equally important to their self-image. "[116] On March 5, Bioplasty withdrew its application from consideration. On March 19 Dow Corning announced that it would be pulling out of the breast implant business. "It was my considered opinion," said CEO McKennon in an interview after the decision, "that there was no viable market for these devices after all of this publicity and after these hearings."[117]

On April 16, 1992, Commissioner Kessler announced the FDA's

decision about implants, a decision that closely followed the recommendations made by the February advisory panel. The resolution, which portrayed the problem as one of uncertainty, maintained the agency's authority over the problem. Implant opponents were able to use the decision, often oversimplified by the media as a statement of harm, to validate their own constructions of reality. Implant proponents, however, once again reconstructed the problem of implants. For plastic surgeons, the problem became one of containing the damage done to their profession and of image rehabilitation. The manufacturers' attention shifted to the management of their legal liability.

In the chapters to come, we will move beyond the staged performances of the public drama by situating the players in their historical and social contexts. We will look at how their interests and values emerged, and thus why they constructed the problem of implants in the ways that they did. But first, let us become acquainted with the central character, perhaps the most mysterious: the breast implant.

2

THE PARADOX OF THE NATURAL

> There is no better beginning to the saga of mammaplasties than "Once Upon a Time" as the history is truly a source of wonder that it is our pleasure to unfold. At its source lies woman's eternal dream—beautiful, firm, and harmoniously proportioned breasts—a dream that has inspired painting, sculpture, and literature.
>
> —J. P. Lalardie and R. Mouly, "History of Mammaplasty"

In 1895 Vincenz Czerny, a prominent Heidelberg surgeon, published a brief case report in a medical journal. It described a woman, reputedly an opera singer, who came to him with interstitial mastitis (an inflammation) and adenofibroma (a benign tumor) in the left breast. Czerny operated to remove the affected tissue. "To avoid asymmetry," he transplanted a lipoma (a benign tumor made up of fat cells) from the woman's flank to her breast, filling in the hollow place surgery had left. One year after the procedure, he wrote, the breast was neither enlarged nor shrunken and it had maintained a good shape.[1]

Czerny, who was not a plastic surgeon, is, nonetheless, the historiographic father of cosmetic breast surgery.[2] His publication marks the beginning of the modern era of breast surgery. It is cited in histories of augmentation, reconstruction, and prophylactic mastectomy. This first spare notice sounds with the preoccupations of the next one hundred years: the aesthetics of the breast, the use of implanted materials, and the restoration of the body after disfiguring surgery. Here, too, are some the hallmarks of the profession: a concern for form and visual harmony and a willingness to innovate.

My discussion of the history of breast implant technology spans three chapters. This chapter looks at the technologies that preceded silicone, covering the period from the late-nineteenth to the mid-twentieth century. Then, chapters 3 and 4 take up silicone. My focus is on the processes of technology development: What prompts the introduction of a new technology? When is innovation accepted? In what circumstances is it rejected? Breast implant technology as a whole developed through a cycle of innovation, modification, and disillusion, a larger pattern that was also repeated with each individual technology. This account also reveals other stories: the role technology played in the establishment of plastic surgery as a profession; how the use of breast implant technology was normalized so that, in many cases, the artificial breast came to be seen as superior to the natural one.

The sources for this history are medical journals, plastic surgery textbooks, symposium proceedings, and short histories of procedures written by and for plastic surgeons. The limitations of these sources include those of chronology and conceptualization. Because of lag time in publication, it is nearly impossible to arrange some of the technologies in strict chronological order. My attempts to construct a chronology also have been frustrated by variations in terminology and a lack of precision, particularly in earlier reports, about diagnostic indications. Conceptually, these sources, because they are based on published accounts, represent only part of the story. I am certain that much more than what is reported here was tried. Publication favors success, but failure is equally important to history. Often the only mention of a technology will be a casual one, incomplete, undocumented. Some of these rumored technologies are included here, particularly those that are repeated over and over again and seem to say something important about the story.

Finally, the greatest limitation of these sources—and thus, of the history I have written—is that they leave a void about the women whose bodies were the proving ground of implant technology. While the medical literature is replete with depictions of their bodies (and, as will be discussed in chapter 4, interpretations of their minds), the women themselves remain silent: headless, handless, stiff figures in black-and-white photographs.

Descriptions of the first cosmetic breast surgeries date back to antiquity. Interestingly, the earliest procedures described were performed to correct gynecomastia—breast growth in the male. Around 650 C.E., Paul of Aegina wrote about surgery to treat the condition: "If, as in women, the breast decline downward, owing perhaps to its magnitude, we make in it two lunated incisions, meeting together at the extremities, so that the smaller may be comprehended by the larger, and dissecting away the intermediate skin, and removing the fat, we use sutures in a like manner."[3] Similar accounts appear in the tenth and eleventh centuries.[4] (Despite all the attention to women's breasts, surgical treatment of gynecomastia has changed little over time. Most modern plastic surgery textbooks include a chapter on techniques for reducing male breasts that look very similar to those used by the ancients.)

Early reports also describe the surgical treatment of hypertrophy, or overly large breasts, which today would be called gigantomastia. One notorious case of hypertrophy was related by Will Durston in 1669: Elisabeth Trevers, a young woman in good health, woke up one morning "and attempted to turn herself in bed, [but] she was not able. . . . Then endeavoring to sit up, the weight of the breasts fastened her to her bed; where she hath layn ever since."[5] Durston claimed to have performed the first reduction mammaplasty on Trevers, but recent scholarship suggests that while he may have made an incision in her breast, he made no attempt to reduce its volume.[6] After Trevers died, Durston did, however, amputate one of her breasts: "Having weighed it, we found it of *Sixty four* pounds weight. Upon the opening of it . . . we could find neither Water, nor Cancerous humors."[7] A barber-surgeon named Hans Schaller, of Ausburg, is credited with the first amputation for hypertrophy, but the date is unknown.

These early operations preceded the introduction of anesthesia and antisepsis/asepsis into widespread use. Surgery, then, was a last resort for only the most dire of conditions. The surgeon had to work quickly and had little time to worry about the aesthetics of the result. By the mid– to late nineteenth century, however, as surgery became safer and surgeons became more skilled, the possible uses for breast surgery expanded, as did the number of potential patients.

In 1882, Theodore Gaillard Thomas, an American surgeon, described a new technique for breast surgery. He used an incision under the breast, where the resulting scar would be hidden, to remove breast masses.[8] The technique soon was adapted for treatment of hypertrophy. In the latter part of the 1890s, French surgeons, contemporaries of Czerny, devised more complex reduction mammaplasties in which entire sections of breast tissue were removed and the breast was reshaped to give it a more pleasing appearance.[9] The early decades of the twentieth century saw even more interest in hypertrophy and the further development of surgical methods to correct it. The innovations were refinements of the resection techniques that had been used for hundreds of years. While the earlier operations had simply removed tissue and closed the skin—often destroying the nipple and leaving the breasts smaller but formless—the new operations attempted to shape the newly reduced breast. In 1909, for example, H. Morestin described a reduction procedure in which tissue was removed from the lower part of the breast and the nipple transposed, through a buttonhole-like incision, to a position at the upper part of the breast. An alternative was to detach the nipple and areola completely, and later, after the breast had been reduced and shaped, reattach them in the proper position.[10]

As techniques became more refined and surgery safer and more successful, the definition of hypertrophy changed. What had been a physical condition (albeit with some emotional consequences) became an aesthetic problem. The hypertrophic breast was no longer only the breast so large it was disabling, but also the breast that was simply too large for preference. With this shift, surgeons began to pay more attention to the aesthetics of the breast: that is, to exploring the ranges of shapes and sizes and describing the ideal. As the bodies of more women came under the scrutiny of the medical "gaze,"[11] the normal increasingly was pathologized, made unnatural.

In 1924 Eugène Briau published an article cataloging breast size and shape among a sample of one hundred working-class women.[12] He described seven types of breast, ranging from the "perfect breast" (*le sein parfait* or *le sein des peintres*) to breasts of abnormal or weird configuration (*les formes anormales ou baroques*). The ideal breast was a perfect hemisphere, firm and elastic, with no folds of skin at its base. Of the one hundred women Briau examined, only thirteen could claim such an ideal form. The others showed varying degrees of sagging, atrophy, or otherwise unaesthetic size or shape. Briau argued that beautiful breasts were a sign of *bon équilibre général* and blamed these defects of appearance on too much hard work, overeating, lack of rest, extended periods of breast-feeding on demand, and poor hygiene.

The same year, another Frenchman, M. Peugniez, explored the source of the aesthetic appeal of the ideal breast.[13] He argued that the beautiful breast was one in which form most closely followed function. The function of the breast was to feed an infant; the most beautiful breast, therefore, was the breast of a young girl—round, firm, full—*sa forme affirmera les promesses de fécondité.* He worried that as Western civilization became less and less concerned with the function of the breast—as bottle-feeding became more widespread—the breasts of future generations would atrophy.

Ironically, it was the breasts of women who had actually fed an infant or two, breasts that *had* affirmed their promise, that most often demonstrated the aesthetic deficiencies that Peugniez and Briau decried. Descriptions of such breasts emphasized their ugliness and imputed a pathological quality: "The withering of the breast and the fall of the teats, in a word, breast prolapse, are the most common characteristics of the degeneration of the female body. . . . this mammary decay sometimes happens prematurely when the female organism is still young. When she still has the beauty and prettiness of her face and the purity and firmness of her body, the fall of the breasts is truly unseemly."[14]

Paraffin injections were one corrective. A hydrocarbon, paraffin exists both in a hard form, which is waxy and moldable, and a soft form, called vaseline, which looks and feels like the product that today goes by that trade name. When heated, both hard and soft paraffin melt into an oily liquid. In 1891, physician J. Leonard Corning reported the use of paraffin oil as a solidifying agent in surgery.[15] By the end of that decade, the Austrian surgeon R. Gersuny suggested that when injected in liquid form, paraffin would solidify in the body and could serve as a subcutaneous prosthesis at sites that needed building up, for example, in the repair of scars and wounds about the face and neck.[16] He used a mixture of one part soft paraffin and three parts olive oil and introduced the mixture into the body in a series of low-volume injections. The theory behind the technique was that the olive oil would be absorbed by the body and disappear, leaving small particles of paraffin around which the body's own connective tissue would grow. The lasting prosthetic function of the method derived not from the paraffin but from the permanence of the connective tissue. In about 1899, Gersuny injected paraffin into a woman's breast, although it is unclear what kind of problem he hoped to remedy.[17] Following Gersuny, an unknown number of surgeons (among them Koch in 1910) injected paraffin and a variety of other substances into the breasts of an unknown number of women.[18]

In a plastic surgery textbook published in 1911, author Frederick Strange Kolle addressed both the advantages and disadvantages of the technique. "The advantage of the Gersuny method over other procedures," Kolle wrote, "is that it can be undertaken practically without pain, that it is quick, bloodless, leaves no scar, and is harmless except under such conditions as will be referred to under a separate heading."[19] The disadvantages he recognized, however, were complications ranging from aesthetic failure to death. Kolle understood these "untoward results" as problems of technique that could be remedied by following proper procedures.

Like Gersuny, Kolle believed that the success of paraffin injections depended on the body's connective tissue response. However, Kolle was hesitant to characterize that response and admitted uncertainty about the ultimate fate of the newly formed connective tissue. He also noted that paraffin could migrate from the site of injection to other areas of the body and that it could cause toxic reactions. Despite recognizing the disadvantages, Kolle remained generally enthusiastic about the paraffin. Contemporaneous accounts, however, show that the technique was highly controversial. In a 1908 article, Chicago surgeon Charles C. Miller described the extremes of opinion: "Some recommend the method warmly, and claim for it tremendous possibilities, while others unfalteringly condemn the technic."[20] Miller's article was a plea that science, not opinion, settle the matter.

It took years, however, for the medical community to reach a consensus. In his 1926 textbook, H. Lyons Hunt called it an "inexcusable practice" and blamed "'beauty doctors' and other such impostors" for its continued use. Hunt's description of the complications of paraffin injection was identical to (in fact, obviously plagiarized from) Kolle's, published fifteen years earlier. However, where Kolle (and Gersuny) had lauded the formation of connective tissue around paraffin molecules and dismissed the accompanying inflammatory response as an insignificant problem of technique, Hunt named such a response "paraffinoma," which he described, quoting Davis, as "a chronic inflammatory process involving skin and adjacent subcutaneous tissues, characterized clinically by the development of reddish-purple, indurated masses, painless, not tender, persistent, not subject to ulceration . . . and ordinarily benign."[21] Hunt claimed to have seen a particularly severe reaction in the breast of the patient Koch injected in 1910: "Within a year the breasts became painful and swellings appeared at the sites of injection. The skin finally broke down at several of these points, and there was a continuous discharge of pus and small pieces of paraffin. . . . The breasts were then very hard with a bluish discoloration." Hunt amputed the woman's breasts. His microscopic examination "showed no large masses of paraffin, but small particles were scattered through the tissues . . . the tissues between showing either fibrous or inflammatory changes."[22]

Hunt argued that paraffinoma was caused by a combination of "individual hypersusceptibility or predisposition" and the introduction of "a non-specific foreign body." He quoted Weidman and Jeffries, who had concluded that "it will finally be shown that the introduction of any foreign body will be followed by an order of tumor reaction, provided the person's tissues are so disposed."[23]

The adverse results of paraffin injections were not limited to the development of benign, localized tumors. In 1930 Karl-Heinrich Krohn published several case reports of women who had had paraffin injected into their breasts.[24] He noted that paraffin traveled in the body by way of the lymph system. This systemic dispersion of paraffin raised questions about absorption, toxicity, irritation, infection, and what he called "connective tissue reaction." The women whose cases he reported not only showed paraffinomas at the site of injection but also exhibited problems of "feverish rheumatism," or polyarthritis, problems that had not appeared until seven to eighteen years after injection. The women's symptoms abated, Krohn wrote, after surgical attempts to remove as much paraffin as possible from the body.

The shift in the medical attitude toward paraffin was due to a fundamental reinterpretation of the results of the injection technique. Such a reinterpretation can be attributed, in part, to a combination of two factors: the years that had elapsed since the first paraffin injections in the 1890s (allowing sufficient time for more severe complications to develop); and the increased number of patients undergoing the procedure (allowing more chances for complications to be seen by individual physicians). Another explanation, however, is that the younger, more sophisticated, generation of physicians was interpreting the same evidence in a different way.

The issues and rhetoric of the paraffin controversy bear a striking resemblance to the issues and rhetoric of the silicone controversy a half century later. As in more recent events, during the paraffin controversy the medical profession followed a similar pattern of blaming technique, then individual susceptibility, and finally the material itself for the adverse effects they were seeing. The experience with paraffin was to live on in the collective memory of the plastic surgery profession. Plastic surgery's own historians blamed the paraffin saga for retarding the progress of the profession, both because the poor outcomes made patients distrustful,[25] and because plastic surgeons, "susceptible to the blandishments of quick results,"[26] relied on paraffin injections rather than developing more technically difficult surgical methods.

Concurrent with the rise and fall of paraffin, the development of new surgical methods continued. The 1920s and 1930s saw a proliferation of publications that introduced new techniques for the correction of atrophy and prolapse. L. Dartiques described several of these surgical procedures. In one, an incision made under the arm was used to tighten the pectoral muscle, thus providing a lift to the breast. In another, the breast was opened and the tissue and underlying muscle manipulated so as to make a rounder shape.[27] In other, more drastic, procedures, surgeons used elaborate patterns of incision and tissue shaping, including even amputation with reattachment of the nipple and areola.[28] Atrophy could be treated by the same techniques as prolapse, but often a simple lift did not create sufficient volume. A 1930 publication by Schwarzmann recommended treating atrophy by inserting glass balls beneath the breast tissue.[29] Later in the decade, echoing Czerny, surgeons began to work with transplants of fat taken from the patient's abdomen or buttocks.[30]

The medical significance of these surgeries lay in the belief that they were more than cosmetic, that they had a broader, curative effect. Surgeons who wrote about these operations argued that the breast was an intimate partner in the normal functioning of the female endocrine system. Hypertrophy, atrophy, and prolapse were signs of systemic malfunction; creating an aesthetically ideal breast restored the body's natural balance. Raymond Passot, who treated atrophy with fat transplants, described patients whose postsurgery breasts continued to grow on their own and eventually regained the capacity for lactation. He accorded credit to the endocrine stimulation provided by the transplant.[31] Madame Noel reported that she had seen miscarriages caused by breast surgery and regulation of irregular menstrual cycles following other "aesthetic" breast procedures.[32] (In contrast, the American physician Lilian Farrar rejected a physiological etiology for breast atrophy and prolapse. She attributed "virginal atrophic prolapsed breasts" to the fashion imperatives of the flapper era in the 1920s. Women bound their breasts to look stylishly flat-chested. The tight bindings caused the breast tissue to atrophy and the breast to fall. Fortunately, Farrar noted, the condition could be remedied by wearing more sensible undergarments.)[33]

The social significance of the breast and of cosmetic breast surgery encompassed other, more subtle, meanings. The 1924 articles by Briau and Peugniez drew on two strong subtexts: a tendency to correlate physical characteristics and medical conditions with class; and an anxiety about racial degeneration and the declining birth rate of the native French. Both of these concerns were located in the historical moment of economic and social upheaval following the First World War—in particular, a challenge to the worldwide class system and a change in the role of women. The breast became a focal symbol for these social subtexts. Indeed, historians of fashion link the flapper's stylishly flat chest and her new social boldness.[34]

For the young profession of plastic surgery, the breast became a site from which to establish the legitimacy of the discipline.

Perhaps the greatest chance for gaining such acceptance lay in the area of reconstructive breast surgery. Breast cancer has been the subject of medical writings for as long as such writings have existed. Ideas about its etiology, and thus beliefs about the appropriate treatment, have paralleled the conceptualization of disease in the history of medicine.[35] Cell theory, the paradigm that revolutionized the practice of medicine and surgery in the mid–nineteenth century, brought a conceptualization of the disease as one of abnormal cells, which could travel to other parts of the body and establish new tumors. American surgeon William Halsted, who in the late 1800s developed the hallmark surgery that still bears his name, believed that breast cancer began as a local condition, but that it could "spread centrifugally along lymphatic pathways by direct extension with the primary growth."[36] The Halsted radical mastectomy removed all breast tissue, the underlying muscles of the chest, the lymph nodes under the arm on the affected side, and a great deal of overlying skin. Success was defined not by long-term survival but by the absence of local recurrence. Such recurrence was interpreted not as a failure of conceptualization but as a sign that the original surgery had not been thorough enough.[37]

The first role for plastic surgery techniques in the treatment of breast cancer was for skin grafts to cover the site of breast amputation and for "salvage" operations in which scarring was reduced or camouflaged.[38] As the Halsted radical became the surgery of choice, the possibilities for a greater role increased.[39] The first reconstructive techniques, developed by European surgeons, used flaps of skin and tissue from the abdomen, back, or remaining breast to build new breasts.[40] In 1903, for example, H. Morestin described an operation in which the remaining breast was cut loose around part of its circumference and stretched across the chest to cover the mastectomy site. The result was a single, oblong breast across the width of the chest, a "breast mound." Morestin noted that some patients, particularly younger women, might become despondent at the appearance of the result, but asserted that a second-stage surgery could be used to split the breast mound and shape two breasts. (There is no record that he himself ever attempted such a second-stage procedure, however.) [41] In 1906, L. Ombrédanne published an article describing a method for mastectomy followed by immediate reconstruction.[42] He amputated the breast and part of the pectoralis major muscle, then cut the pectoralis minor, twisting it into a flap which he then reattached, forming a breast-shaped protuberance. The new breast was covered with a flap of skin and tissue from the abdomen.

That these first reconstructive procedures were described by European surgeons is not a coincidence. Americans were strict followers of Halsted, who was opposed to reconstructive surgery after mastectomy. His opposition was based upon the belief that surgery itself could spread cancer, either by loosening tumor cells or through tumor "seeding" accomplished with a contaminated knife. In addition, he feared that a reconstructed breast might hide a local recurrence.[43]

The dominance of the Halsted philosophy in the United States made surgeons who performed mastectomies very reluctant to recommend reconstruction to their patients. Despite the resistance of the larger medical community, however, American plastic surgeons continued to work on developing new reconstructive methods. In fact, the introduction of each new implant technology was accompanied by speculation about its utility for reconstruction. Clearly, plastic surgeons saw restoration of the breast lost to surgery as a great professional and technical challenge.

The rage for the flat-chested look in the 1920s no doubt contributed to the lack of attention to small breasts. Allusions to procedures for the enlargement of small breasts (not atrophic or prolapsed breasts) are indirect. Otto Frisch described a woman who "complained about her childish appearance and wished to appear somewhat more stately."[44] Citing the inadequacies of fat transplants, he declined to recommend surgery to the patient. Koch's 1910 use of paraffin injections for breast enlargement is cited in the contemporary literature, but Passot's statement—"I mention it only to condemn it"[45]—is typical. The late 1930s saw a flurry of publications about the psychological effects of "abnormal" breasts, and, embedded within the discussion of psychology, an argument for the legitimacy of cosmetic work on small breasts.

In a paper presented at the 1935 meeting of the recently

founded Society of Plastic and Reconstructive Surgery, H. O. Bames described three types of "anomalous" breasts: prolapsed, abnormally large and pendulous, and abnormally small. These "anomalies" were deemed problematic not because they caused a disturbance in function—as Peugniez had predicted a decade earlier, "modern methods of infant feeding" had vitiated the need for breasts that could lactate—but because there was something unnatural, and thus disturbing, about such "anomalies of form." "What woman with abnormal breast development," Bames wrote, "does not look with envy upon the well formed figure of another woman, and rightly so, for is not a well formed figure the outward manifestation of normal health and development, and thus representative of an ideal to which we all subscribe?"[46]

For the treatment of overly large and pendulous breasts, Bames described a reduction mammaplasty with nipple transposition. He rejected the possibility of a surgical solution to small breasts, however, noting the inadequacies of fat transplants and saying these anomalies were "a problem for endocrine rather than surgical correction."[47]

In a textbook published in 1942, author Max Thorek, a Chicago surgeon, described the normal breast, with anatomic exactitude, as "two hemispherical projections upon the thoracic cage, one on either side of the sternum. They extend from the lateral border (margin) of the sternum to the axillary border and from the level of the second to that of the sixth rib. The outline of the breast is elliptical. The horizontal diameter measures from 10 to 12 cm. and is about 1 cm. more than the vertical dimension."[48]

Thorek's textbook presented a broad summary of what was currently known, and thought to be important, about the female breast. It included chapters on anatomy and histology, physiology, historical and comparative analyses, folklore of the breast, and racial characteristics. It covered the etiology, symptoms, course, and complications of hypertrophy; surgical techniques of reduction mammaplasty; and diseases of the breast and nipple. In addition, Thorek devoted a chapter to a discussion of small breasts, a condition he called hypomastia. Although less than 10 percent of the total number of pages in the book, Thorek's disquisition on small breasts included a classification scheme, a brief discussion of etiology, and a review of the treatment methods currently available.

Thorek described three degrees of hypomastia: first degree, or the "nipple breast"; second degree, in which the breast structure was slight; and third degree, in which the breast was disproportionately small for the woman's build. He noted two varieties of hypomastia. In the first, there was a total lack of fatty tissue, but the nipple and gland structure were normal (this was the so-called nipple breast). In the "infantile breast," however, all structures, including the nipple, were undeveloped. Hypomastia's causes were either congenital, hormonal, or acquired (through surgery or trauma). Treatment possibilities included hormones (estrogens by mouth, injection, or in the form of a topical ointment); paraffin injections ("[the technique] is mentioned here because of its historic interest . . . injections of paraffin should never be given"); prostheses (specifically, Schwarzmann's glass balls); and finally, surgical measures like fat transplantation.

The state of the art in the surgical treatment of hypomastia was evidenced in a 1944 article by surgeon Morton Berson.[49] Berson took fat and skin from the buttocks and inserted it into a pocket made between the tissue and underlying muscle of the breast. The addition of dermal tissue to the long-used technique of fat transplants was a direct response to the well-recognized inadequacies of the earlier technique—the tendency of fat transplants to liquefy and be resorbed by the body. The derma-fat-fascia combination, Berson argued, "is more easily handled, gives better contour to the breast, and produces a firmer filler substance."[50]

Innovation in the 1930s and early 1940s had less to do with the technology of breast plastic surgery than with the establishment of a broader role and context for that surgery. The problem of the breast was now articulated as a complex web of social, cultural, and psychological factors. (See chapter 4.) At the center of that web was the patient, who was construed as a supplicant. It was the professional obligation of the plastic surgeon to respond to her demand. The available technologies, however, fell far short of satisfying the obligations incurred by the pathologization of so many different breasts. In the confluence of several war-influenced trends, surgeons turned their skill and knowledge to developing new technologies that would allow them to fulfill their obligations by better meeting the demands of their patients. As technology improved, plastic surgeons increased their promotional efforts; thus demand also grew.

Postwar innovation in reconstructive technique was prompted by the changing demographics of mastectomy patients. The educational messages of early detection and early treatment, suggested by the cell theory of breast cancer, were working. More women were having more mastectomies at younger ages and were living longer after surgery.[51] In the late 1940s, plastic surgeon Harold I. Harris noted the changing population and longer survival time of mastectomy patients in an article in which he described a new, multistage reconstruction procedure using tissue from the remaining breast.[52]

As Harris suggested, some of the improvement in survival time after mastectomy was due to operations that turned out to be for benign conditions.[53] Publications back to Czerny show that partial resections for conditions described as "chronic mastitis" or "fibroadenoma" had been common for the last half century. In articles published early in the twentieth century, however, it is unclear whether surgery was indicated because of the abnormality itself or because these benign conditions were thought to presage the development of a malignancy.

The first prophylactic mastectomy—surgery with an explicitly preventive aim—was described in 1917 by Willard Bartlett.[54] He argued that what mattered to women was not the function of the breast but its form. If surgeons could provide an operation that would remove the vulnerable glands and tissue of the breast, but maintain its appearance, women would consent to have surgery early, before benign tumors could turn malignant. Bartlett described a procedure in which he used cautery "to shell out the entire gland-bearing area" of the breast, then inserted fat, taken from another site on the patient's body, "to fill out the defect."[55]

In 1951, surgeons Carl Rice and J. H. Strickler cited Bartlett in an article describing a similar "shelling" procedure.[56] After making an incision in the submammary fold, they dissected the breast tissue away from the pectoralis muscle. The dissection was extensive enough to allow the breast to be turned "inside-out" for examination. The surgeons then removed the tumor and some or most of the surrounding breast tissue. The operation left what they called a "breast capsule" of skin and subcutaneous fat which was, they argued, adequate to maintain an aesthetic appearance.

The outlines of the subsequent history of prophylactic subcutaneous mastectomy are apparent in these early reports. That history has three important strands: the belief that the benign conditions subsumed under the appellation "fibrocystic disease" were premalignant; the argument about the extent of surgery needed to provide prophylaxis; and the development of improved methods for reconstruction.

In the 1950s, there was a transition from techniques that used the body's own tissues to new technologies of the artificial. Postwar interest in more complex surgical procedures and advances in the synthesis of man-made materials that had occurred during and after the war converged in a research effort to develop biocompatible synthetic materials for surgery.[57] Reports of attempts to use nylon[58] and Plexiglas[59] for breast plastic surgery appeared in the literature. More widespread was the use of polymerized synthetic sponges, including Ivalon (polyvinyl alcohol), Polystan (polyethylene), Etheron (diisocynate polyester), polyurethane ester, silicone sponge, and Teflon.

The introduction of these materials was met with both excitement and controversy. Their ease of use and good (initial) results led an increased number of surgeons to perform more breast surgeries. However, increased application drew attention to the problematic nature of the body's response to the materials. By the end of the decade, a debate raged among plastic surgeons over the natural and the artificial. In an echo of the paraffin controversy, surgeons struggled to understand the meaning of the body's response to artificial materials, and asked questions about the possible long-term and systemic effects of implants.

Harbingers of the shift to artificial materials came in a series of articles describing laboratory and animal studies of the various polymers. These studies examined the chemical and structural characteristics of the new materials, focusing on their potential for use as long-term implants. John Grindlay and John Waugh, two scientists who worked with polyvinyl alcohol, entitled one of their articles "Plastic Sponge Which Acts as a Framework for Living Tissue,"[60] thus clearly drawing a distinction between the artificial material of the implant and the body's own tissues. Yet their work showed a tendency to elide that difference. Indeed, they reported that polyvinyl alcohol not only was "stable and biologically inert" but was able to "become part of the structure of living tissue."[61]

The statement was based on a series of studies in which sponge implants of various sizes were inserted into dogs. After a period of eighteen months, the dogs showed no signs of any adverse effects. When removed for examination, the sponges were found to have been enveloped by "a thin and delicate fibrous semitransparent capsule" containing "blood vessels and fibrous connective tissue." This capsule clung both to the implanted sponge and to the surrounding tissue of the body, giving the sponge "the appearance of a strange but somehow normal appearing organ."[62]

Their confidence that this "strange organ" could "become part of the structure of living tissue" was grounded in evidence of both structural and physiological compatibility. Grindlay and Waugh noted that the connective tissue capsule was useful for fixation of the implant: it would stay where it was placed. They interpreted the capsule as a sign of the body's acceptance of the implant—indeed, of its naturalness. "[The] affinity of polyvinyl sponge for water . . . and the fact that the tissue cells live in water environment," they wrote, "give the sponge a kinship with living tissue. Perhaps tissue fails to recognize polyvinyl sponge as a foreign body, because tissue fluids enter it. Cells follow fluid, and what was inert becomes living."[63]

The work of other scientists, however, tended to emphasize the continuing artificiality of such synthetic sponge implants. A 1952 study of mice implanted with polyethylene found similar capsular development but also reported a high incidence of cancer among the implanted mice.[64] Although the authors indicated that what was true for mice might not be so for human beings, they did attribute the tumors to some chemical property of the material, suggesting, at the very least, some caution.

That breast plastic surgery was among the envisioned uses for synthetic sponges was apparent from the first. (Grindlay and Waugh even cut one test sponge into the shape of a miniature breast and implanted it beneath a dog's nipple.) Existing technologies had proved inadequate in use: fat transplants shrank; paraffin was a disaster. As soon as synthetic sponges became available, plastic surgeons began to use them in clinical practice.[65] The history of synthetic sponge implants is best told through the stories of two men: W. John Pangman and Robert Alan Franklyn.

W. John Pangman, a plastic surgeon who practiced in California and Texas from the 1940s until the late 1960s or early 1970s, has become identified as the profession's pioneer in the use of synthetic sponges in breast surgery. Pangman used Ivalon (the trade name for polyvinyl alcohol), which he described as "a white sponge material with the appearance of white bread," to fashion a series of implants.[66] His first design was an oval-shaped piece of Ivalon, convex on one side and flat on the other, measuring about four and a half by two and a half inches. Pangman implanted the "breast prosthesis" by making an incision in the fold under the breast, using a blunt instrument to detach the breast tissue from the underlying muscle, and inserting the implant into the space created.

Pangman praised the cosmetic results of his surgeries and reported that a survey of his patients had found a high level of satisfaction. These early implants, however, tended to cause problems with infection and drainage, often leading to their rejection by the body. In response, Pangman improved sterilization procedures and took to irrigating the breast pocket with antibiotic solutions. A more intractable problem was hardening and shrinkage of the breast. As Grindlay and Waugh had described, the body responded to the presence of a polyvinyl alcohol sponge by forming a capsule of connective tissue around the implant. Over a period of months, Pangman reported, the Ivalon implant would be "invaded" by a fibrous tissue ingrowth from this capsule, causing the prosthesis to shrink and harden. (He estimated the degree of shrinkage at 25 to 75 percent of original volume.)

Pangman tinkered with his design, trying to reduce some of the problems he found in clinical use. He developed a new "compound prosthesis" that used an inner core of foam encased in a layer of polyethylene plastic and sewn inside another thin "shell" of polyvinyl foam. As he explained, the connective tissue would grow into the outer shell of foam, allowing fixation of the implant to the chest wall, but the polyethylene film would act as a barrier against further ingrowth, preventing shrinkage and hardening. In a 1955 report on the compound prosthesis, Pangman claimed success, reporting that "shrinkage has been from zero to twenty percent. The breasts are usually firmer than normal but not objectionable and some implants can not be detected by palpation at all."[67] (Perhaps belying this apparent satisfaction, Pangman later introduced further refinements to his implant, including heat sealing the foam core and replacing the polyvinyl alcohol foam shell with one fabricated from polyester.)

In a 1979 interview with plastic surgeon Garry Brody, an aged and ailing Pangman located the impetus for his experimentation with Ivalon in the "crying need" expressed by his female patients.[68] Indeed, in his articles, Pangman promoted the use of implants for augmentation, reconstruction, and prophylactic mastectomy. He focused on the latter two indications (although it seems clear that most of his surgeries were for the purpose of augmentation) and warned against "the promiscuous use to build up the small breast." Implantation should be performed only for selected cases and "only by those well trained in the surgery, anatomy, and pathology of the breast."[69]

Despite the care with which he presented his work, Pangman faced strong personal and professional opposition from his colleagues. (In the 1979 interview he described the criticism leveled against him as "brutal.") A 1991 plastic surgery textbook analyzed this reaction: "Common dogma dictated that the body poorly tolerated foreign materials. 'Real' plastic surgeons used autogenous tissue only. . . . Use of a foreign object in the breast was an even greater anathema, especially for 'frivolous' cosmetic enlargement. Prior to the 1950s, aesthetic surgery lacked the status and respect associated with reconstructive procedures. It was carried out by most, but admitted to only by a few brave souls who suffered denunciation by the Establishment."[70]

Robert Alan Franklyn attracted even greater animus. Franklyn practiced in Los Angeles in the 1950s. Although never credentialed as a plastic surgeon, in the early 1950s he began performing breast augmentation procedures using a mysterious substance called "Surgifoam," which may or may not have been polyvinyl alcohol sponge. (Pangman's early publications seem to have been prompted in part by a priority dispute with Franklyn, whom he called the

"Hollywood Physician," and probably had in mind when he warned of unqualified surgeons performing "promiscuous" enlargements.)

Franklyn was known for several things: an early article that described what might have been the first implant design to use silicone;[71] a list of ideal qualities to which materials used in breast plastic surgery should conform;[72] and the Breast Quotient (BQ), a formula for determining the degree of hypomastia:

$$BQ = \frac{\text{hip measurement x frontal chest measurement}}{\text{height}}$$

A BQ of 100 indicated severe flat-chestedness.

Franklyn's bad reputation was a result of his practice of speaking directly to the public, even authoring articles for the popular press. In 1953, Franklyn was investigated by the American Medical Association (AMA). The investigation was prompted by an article in *Pageant*, a popular magazine, in which Franklin asserted that "four million women suffer from micromastia, another 10 million from ptosis [sagging breasts]" and described a "relatively simple, 25–minute operation" to correct the conditions. In the procedure, Franklyn wrote, an "inert soft plastic foam substance—actually a nylon-like resilient material impregnated with penicillin" is inserted into the breast, where it "adds volume, forcing the chest out, but . . . does not interfere with the complicated glandular structure."[73]

In a report of the AMA's investigation published in *JAMA*, Franklyn was made the subject of ridicule: "As well can be imagined, this glowing invitation to pectoral pulchritude has caught the fancy of a good many who feel they are among the 4 million. . . . one wonders how one operator is going to manage this singlehanded. He can take care of only 7,280 if he works 10 hours a day, 7 days a week, for a solid year. One needs to extend the figures only one step further to see what frustration may face the overwhelming number of those left out."[74] And also, more seriously, he was accused of unprofessional conduct and irresponsibility: "It is strange indeed that a physician would permit publication of the details of a surgical procedure in a popular magazine when such procedure does not appear to have had the benefit of exhaustive tests on animals to establish its safety and efficacy."[75]

Although the investigators were uncertain about the exact composition of Franklyn's "Surgifoam," they assumed it was similar to polyvinyl alcohol and cited the work of Grindlay and Waugh in cautioning against "at least the prospect of a gristle-like development in the implant which is apt not to be what most ladies would like to carry around with them."[76]

That this judgment came from the AMA—the voice of the medical establishment—was ominous. Like Pangman, Franklyn was criticized for his use of an artificial and untested material, a criticism that implied he was unscientific. Beyond that, however, investigators seemed most concerned about Franklyn's direct address to the public, a practice that could be construed as advertising. These criticisms implied a threat to the entire discipline of plastic surgery, for not only were Ivalon and other synthetic sponges becoming the standard of the profession, but—as plastic surgeons were learning—continued development of the technology required its constant promotion to likely consumers.

While Pangman's status as an advocate for implants brought him much contemporary disapprobation, the historiography of plastic surgery has transformed him into Czerny's heir. In the mid-1970s, Pangman's reputation began to be rehabilitated. One textbook argued that "he was a controversial and unappreciated figure. . . . Yet his historical significance cannot be denied or diminished."[77] As with Czerny, Pangman's great contribution was to innovate and, beyond that, to modify his own technology. His legacy was to establish the idea that synthetic materials could be used successfully. In contrast, Franklyn has become the dark twin, an exemplar of accusations that continue to haunt the profession. For him, there has been no vindication.

In 1955, at the First International Congress of Plastic Surgery, two of the five papers on the topic of breast plastic surgery were devoted to favorable reports on the use of synthetic sponge materials in the treatment of hypomastia.[78] As more prostheses were implanted, however, more plastic surgeons witnessed firsthand the subsequent shrinkage and hardening of the breast and the occurrence of postoperative infection. (And more performed the "difficult and bloody" procedure needed to remove a firmly fixed Ivalon implant from a breast pocket.)[79] Some surgeons were moved to disavow the use of polyvinyl alcohol (and polyvinyl chloride) for use as breast implants. "It is realized that nature abhors a foreign body," one group of authors wrote.[80] A 1956 review of mammaplastic techniques noted that the use of Ivalon "foreign body prostheses" was a "very controversial subject," claiming that "the very great majority of recognized plastic surgeons are opposed to its use not only on the grounds that it is experimental, not scientifically tested, not permanent and gives the breast an unnatural consistency but primarily that the interposition of a foreign body under the susceptible glandular tissue of the breast may lead to serious degenerative changes in the foreseeable future."[81]

These criticisms from inside the profession emphasized the "unnatural" qualities and results of Ivalon and focused on the likely occurrence of local adverse effects. Plastic surgeons' objections seemed to be directed against the specific material and not the idea of implants made of artificial materials. Outsiders, however, attacked the whole idea of artificial implants, suggesting the possibility of systemic harm.

William Kiskadden, a Los Angeles physician and a particular nemesis of Pangman, labeled plastic surgeons' belief that "the soft sponge would continue its resilience and give to the breast a relatively normal consistency" as "wishful thinking." Instead, he noted, "the natural consistency of the breast has been lost and all that has been gained . . . is a somewhat larger breast mass."[82] "It therefore seems highly dangerous, unscientific and without valid clinical reasoning or application," he concluded, "to introduce a foreign plastic material whose immediate results are so indifferent and whose long term outlook is filled with the potential danger of cancer."[83]

The possibility that sponge implants might be carcinogenic was raised again in a report by W. C. Hueper, head of the National Cancer Institute's Environmental Cancer Section. His studies of Ivalon, polyethylene, and other synthetics in the late 1950s strongly suggested that the materials could cause cancer in rats. He cautioned against clinical use of the materials until the issue of carcinogenesis was better understood.[84] Another group of researchers, whose work with polyvinyl sponge had had similar results, was loathe to extrapolate to humans but noted that, as in humans, in rats the first sign of a reaction was the formation of a capsule of connective tissue around the implant. Their advice to plastic surgeons—"to use sheets of sponge as thin as possible . . . and to remove them . . . when their purpose has been served"[85]—was a clear warning against using the material for breast surgery.

These critics emphasized the gulf between plastic surgery's process of technology development and mainstream science. Their arguments against sponge prostheses marshaled the results of animal testing, an implicit acceptance of the idea that findings in animals could—and should—be extrapolated to humans. In addition, they conceptualized local adverse effects and systemic harm as a continuum. By contrast, plastic surgeons who were proponents of sponge technology rejected the relevance of data based on animal studies. Instead, they argued that an understanding of the effects in humans could be gathered only by implantation in humans. If adverse effects were detected, that was not a reason to reject the technology; rather, such effects might be the result of poor surgical technique, minor features of design, or even the individual patient's physiology. Proponents also rejected the idea of a continuum between local and systemic adverse effects. They separated the two, categorizing "unnaturalness" as a local problem of poor result. It was clear that these plastic surgeons viewed the potential of the artificial as superior to the known qualities of the natural.

The reactions of synthetic sponge proponents to criticism of the technology reflected their assumptions. Their response to the question of carcinogenesis was to conduct surveys of plastic surgeons who used the sponges to determine the incidence of cancer among women who had been implanted with them. (Implicit in this tactic was the argument that plastic surgeons possessed the scientific acumen to perform such studies.) One such survey, sent to all certified plastic surgeons in the United States,[86] found no cases of cancer in the ten years covered by the survey.[87] The authors of another survey, recognizing that plastic surgeons were unlikely to be the ones to diagnose postimplantation breast cancer, or even to know of the diagnosis, confined their inquiry to cancers discovered during surgery.[88] Some plastic surgeons theorized that the risk of cancer, if such a risk even existed, was just as likely to be the result of surgery as it was the implantation of a foreign body. Thus the authors of the latter survey included cases of breast reduction and gynecomastia in their survey.

Another response was to develop a more rigorous approach to the psychological indications for breast plastic surgery. While plastic surgeons had, in the past, reported psychological indications and results in an anecdotal, casual fashion, the new emphasis on risk called for a more "scientific" explication of benefit. If the implantation of synthetic materials was suspected of being dangerous, satisfying a purely cosmetic desire was not enough of a benefit. It was here, then, that "hypomastia" shifted from a purely descriptive term to a diagnostic term, one suggestive of a particular psychological syndrome. (This shift is explored at length in chapter 4.)

The response to reports of local problems (hardening, shrinkage, "unnaturalness") was to blame specific features of the technology and to attempt to modify them or to design innovations to correct the shortcomings—a strategy exemplified by Pangman's attempt to build a better breast implant. Several plastic surgeons published articles in which they listed the "ideal qualities" of the envisioned ideal implant: (1) it would not be physically modified by tissue fluids; (2) it would be chemically inert; (3) it would not excite an inflammatory or foreign body response; (4) it would be noncarcinogenic; (5) it would not provoke an allergic response; (6) it would resist mechanical strain; (7) it could be fabricated in the form required; and (8) it could be sterilized easily.[89] These "ideal qualities" centered on both ease of use for the surgeon and the need to find an artificial material "natural" enough that the body would accept it. At first, other sponge materials would seem to satisfy these criteria, but eventually all would be found lacking.

The debate over the use of artificial materials revealed an epistemological split among plastic surgeons. Those who resisted the use of synthetics subscribed to the ancient principle that "like should replace like."[90] Synthetics were an aesthetic and moral abomination, an insult to the artistry and technical skill of the surgeon. Those enthusiastic about synthetic materials, however, pointed out that their results were often cosmetically superior to those that used autogenous tissue alone. "Chemistry is a very old and advanced science," they argued. "Let us not close our minds to its possible contributions in the matter of new compounds with potential clinical applications."[91]

The language of the debate suggests a blurring of the distinction between surgeon and patient, a metaphor in which the body's acceptance of surgically implanted synthetics and the profession's acceptance of the materials were equated: "Since the surgeon must live with his implant nearly as intimately as his patient, his decision to use unnatural materials is not made lightly or implemented casually."[92]

The argument over the artificial contained elements of generational and class conflict, of tension between old-school reconstructive surgeons—who performed a few intricate procedures on an elite (by virtue of their degree of disfigurement or disability) clientele—and the new generation of surgeons. This new generation, armed with the technological developments of the postwar era, sought to make cosmetic improvement available to the masses. The debate turned around money and power and the direction the profession would take in the future. As an observer would write years later, it was only public demand that settled the debate: "Were it not for this great pressure, surgeons would still be arguing about using God's own materials versus man-made substances."[93] Arguably, it was the promotional practices of the new generation that helped to create this demand.

The debate also engaged the meaning of the breast. In the 1950s, fashions in women's clothing and bodies changed. The new ideal was voluptuous—an hourglass figure of breasts and hips—the body of Jane Russell, of Marilyn Monroe. This shift to an exaggerated womanliness has been linked to the postwar political economy in which women were urged to leave their wartime jobs to returned soldiers and to devote themselves to motherhood and the home.[94] One meaning of the breast, then, lay in the desire created by the social pressures of advertising and fashion, the forces of the new consumer culture. Plastic surgeons acknowledged these forces in their recognition of a "bosom conscious society." The breast represented a healthy, active lifestyle. A lifestyle that, as in some advertiser's dream, was a constant round of tennis, sunbathing, and semiformal dances—all activities that drew attention to the breast. Often plastic surgeons characterized the need for implants as a problem of withdrawal from such activities, a withdrawal that could be miraculously reversed by surgery.

The breast that revealed itself to plastic surgeons was an

"amorphous mass of fat and glandular tissue"[95] in need of the surgeon's skill. Its function was ignored, save perhaps in brief mentions of breast-feeding as a cause of unaesthetic atrophy. (Yet another case where the artificial, bottle-feeding, might be found to be superior to the natural.) Breast tissue was problematic—too much or too little. The "susceptible glandular tissue of the breast" threatened to turn cancerous.

The natural and the artificial would take turns and reversals as the breast and technology merged. In the end there was a paradox: the surgeons' rhetoric argued for the inadequacy of the natural. Variation was pathologized. Yet always the aim of using the artificial was to re-create an ideal breast, an ideal modeled on the natural. The challenge was to find the artificial material that could replicate this ideal. With silicone, plastic surgeons believed that they had found that material.

3

A PLEASING ENLARGEMENT FROM WITHIN

The movie *Breast Men,* aired on HBO in 1997, offers one portrayal of the silicone breast implant story. In this account, which the opening credits tell us is based on fact "only slightly augmented," David Schwimmer plays a young plastic surgical resident whose voyeurism—he peeps into the bedroom of an apparently voluptuous neighbor and is disappointed when her bedtime ritual reveals that she wears falsies—inspires him to create a breast implant. He brings the prototype to his attending surgeon, played by actor Chris Cooper. Mindful of the disaster of sponge implants, Cooper is dismissive. Later, however, after a party at which he is taunted by colleagues who are developing lifesaving surgical techniques, the older man relents. The two convince a woman to have the device implanted, fine-tune their design, sell the idea to Dow Corning, and find themselves awash in bright sherbet-colored swinging sixties success. Alas, this is a morality tale, of sorts: they grow too successful, too arrogant. Their relationship disintegrates. The partnership dissolves. Cooper dies alone, embittered; the breast implant has brought him riches but not the respect of his peers. Schwimmer succumbs to excess. He becomes the house surgeon first for local strip clubs and then for *Playboy, Hustler,* Hollywood. Cocaine turns his world frantic. He loses everything, then, after news of the possible dangers of silicone hits the headlines, makes another fortune removing implants from panicked women. His sudden death in a violent car crash seems an abrupt justice.

As the inadequacies of synthetic sponges grew more apparent, plastic surgeons turned their efforts to finding alternative technologies. Silicone was a prime candidate early on. In 1953, a "preliminary report" about silicone, noting that all studies performed thus far had found that the material was "inert," suggested that it might fill "one of the greatest needs in plastic surgery, that of a permanent subcutaneous prosthesis."[1] A 1959 experiment showed that when firm, closed cell silicone sponges were implanted into dogs, they were surrounded by a thin capsule of fibrous tissue but did not develop any tissue ingrowth.[2] At the 1960 meeting of the American Association of Plastic Surgeons, a presentation on silicone was quite favorable. The researchers reported that silicone seemed to fulfill the list of "ideal qualities" that plastic surgeons had envisioned for implants. In addition, the material was easily available and stored and was inexpensive. However, they noted that because silicone resisted the ingrowth of fibrous tissue, implants tended to slip out of position. They also warned that silicone had not been in use long enough for its long-term effects to be observed and that much investigative work remained to be done.[3]

Although none of these reports were focused on the use of silicone for breast surgery, the material was quickly adapted for such purposes. At the 1962 meeting of the American Society of Plastic and Reconstructive Surgeons, Benjamin Edwards reported his use of a silicone sponge implant covered with a thin layer of Teflon felt. (The felt, he explained, provided a textured surface to which the fibrous capsule could adhere, thus solving the fixation problem.) Edwards himself had used the implant in twenty-one women over the past three years, and other surgeons had implanted them in two dozen more patients. The general results "have been satisfactory to both patients and surgeon," he said, although most cases "do not necessarily look as natural [as the best results]."[4]

Silicone is a synthetic compound, a polymer, made up of linked chains of the monomer dimethyl siloxane. This monomer marries a methyl group with oxygen and silicon, a nonmetallic element.

This marriage results in a highly desirable mixture of qualities: from its organic component, silicone exhibits flexibility; from the inorganic, stability. Different polymerization processes result in chains of different lengths, which yield products that vary in viscosity—from liquid to semisolid gel to firm rubber. Silicone polymers are highly resistant to heat, oxidation, and chemical degradation. They repel water and, when applied to metal surfaces, decrease resistivity.[5]

The history of the silicones began in the late nineteenth century with British chemist Frederick S. Kipping, who coined the term and performed the first production experiments.[6] The industrial use of silicone started in the years after the First World War when the focus of the chemical industry was on the development of plastic compounds, man-made materials that would be cheaper to produce and more versatile than the natural products already available. The Corning Glass Company, an American concern, was one business to recognize that the future lay in such synthetic materials.[7] In 1930, the company hired its first organic chemist, whose mission was to pick up where Kipping had left off and develop useful products using silicone chemistry. Through its manufacture of glass, Corning had an abundance of available silicon but needed a source of organic chemicals. It formed a working partnership with the Dow Chemical Company. The first silicone products produced by the partners included resins used as mortar for glass brick construction, insulation materials for electrical wiring, lubricating oils for machinery, antifoaming agents that could be added to engine oil, and an ignition sealing compound of rubbery texture that later was repackaged and sold as a children's toy: Silly Putty.

The demand for silicone products soared during the Second World War. Greatest need was for the liquid silicone used as a lubricant in airplane engines. To meet military demand, Corning Glass and Dow Chemical formed a jointly held company that would produce silicone products exclusively. The new company was called Dow Corning and was located in Midland, Michigan.

Postwar peace brought a sharp decline in business for Dow Corning. The company regrouped, focusing on one objective: "Find something people will buy!"[8] At first, Dow Corning emphasized new uses for the products it already manufactured. For example, it marketed silicone liquid as a lubricant for the molds used by tire makers and commercial bread bakeries. Soon, however, the company formed a new market research department which, in conjunction with the advertising department, worked to publicize the properties of silicone and promote its use in new products. One of the markets with which Dow Corning made contact was the medical industry. In 1950, the first in a series of articles about the possible medical uses of silicone appeared in a medical journal. The article's authors concluded that "from our preliminary studies, this new group of silicone fluids . . . has created a logical basis for believing they hold great possibilities in the medical, dental, pharmacological and allied fields."[9] Among the suggested uses for silicone: a gel-like suspension medium for drug delivery (e.g., contraceptive jelly), a waterproofing agent for dressing wounds, an antibloating treatment for cattle, and a lubricant for laboratory equipment. The first medical device actually produced by Dow Corning was a silicone rubber shunt for the treatment of hydrocephalus. By the early 1950s, silicone was being promoted as the ideal material for implantable prostheses.

In 1963, at the annual meeting of the American Otorhinologic Society for Plastic Surgery, Silas Braley, director of the Dow Corning Center for Aid to Medical Research, praised the silicones for their stability and lack of reactivity in the human body. "The body can almost be said to be unaware that they are present," he said. He described the many surgical uses to which silicone already had been applied, including for reconstructive surgery of the face and head, and hinted at the newly developed silicone gel breast implant—"a still softer and more natural implantable breast"—an innovation that would not be introduced officially for several more months. Braley offered two cautions about the use of silicone: first, that while silicones seemingly produce little reaction, "they are, nevertheless, foreign bodies, and continued evaluation should be made before efficacy can be vouched for unequivocably"; and second, that surgeons should take great care to work only with those silicones rated by Dow Corning as being medical grade.[10]

Comparison with older technologies was explicit in these reports, as was the conclusion that the future lay with silicone. The process of technology development seemed designed to assess the likelihood of silicone causing the same kinds of problems as earlier synthetic materials. Silicone was implanted first in dogs, whose local reactions were observed for a short time. It underwent some chemical and mechanical assessment—immersion in solutions of acid or base, exposure to different temperatures. Human testing followed quickly. (Since many of these "researchers" were practicing surgeons who used new devices on their patients, it is questionable whether this early use can really be considered testing.) Triumphant reports of the "inertness" of silicone were based on an assessment of the fate of the material itself. Silicone did not disintegrate in the body. It remained unchanged under different chemical and temperature conditions. "Nonreactivity" referred to the lack of any apparent adverse response by the body. There was no fluid buildup, no infection, no tissue ingrowth.

Advertisements for silicone products emphasized these qualities and proclaimed silicone something better than natural. A 1964 ad for silicone sponge that appeared in *Plastic and Reconstructive Surgery*, for example, used the tag line "nature couldn't produce this inert 'material.'" Its text explained that "nature, usually so adept at meeting man's needs, just couldn't produce a flexible type material which was inert, heat and cold resistant, and non-reactive to tissue." What nature couldn't do, silicone could.

The silicone gel breast implant was unveiled at the Third International Congress of Plastic Surgery, held in Washington, D.C., in 1963. In a paper titled "Augmentation Mammaplasty: A New 'Natural Feel' Prosthesis," Thomas Cronin (a plastic surgeon from Houston, Texas) and Frank Gerow (his resident at the time the implant was developed) recollected the history of their innovation.

> Many women with limited development of the breast are extremely sensitive about it, apparently feeling that they are less womanly and therefore, less attractive. While most such women are satisfied, or at least put up with "falsies," probably all of them would be happier if, somehow, they could have a pleasing enlargement from within.
>
> This natural desire has resulted in many attempts during the last ten or twelve years to achieve enlargement. . . . while many of these have produced results that might be pleasing to the eyes, all have been characterized by hardness, sometimes rock-like and at all times unnaturally firm.
>
> For many years, the senior author felt that an implantable

> breast prosthesis should resemble the soft, somewhat fluid-like movable nature of the normal breast. . . . It was not until 1960, that a promising material (silicone rubber) for the construction of the device was learned of. With the invaluable help of Dr. Gerow and the cooperation of Dow Corning Center for Aid to Medical Research, work was started.[11]

Soon, an observer wrote, "Gerow returned as he had gone, in his rumpled suit, with necktie flying, ever present cigarette, and paper coffee cup clutched in one hand. But in the other hand he held the first man-made pair of Silastic gel breast prostheses! The country was in a breast-conscious, affluent, youth-concentrated mood, and Gerow laid on Cronin's desk the answer to the demand of the times."[12]

The answer was a silicone rubber (trade name, Silastic) implant filled with a firm silicone gel. The posterior side of the implant was covered with a sheet of Dacron fabric, a feature Cronin and Gerow believed was necessary to facilitate fixation to the chest wall: "Fibrous tissue [will] invade this backing, taking the weight off the overlying skin, thereby preventing the subsequent development of ptosis of the breast."[13] Cronin and Gerow had settled on the silicone gel filling (developed by Dow Corning for use as electrical insulation) after testing several other substances. ("The well developed normal human breast has many of the characteristics of a fluid filled flexible container," they noted.) The gel was superior, they wrote, because it "is very cohesive, that is, it sticks to itself and so is essentially leak proof, even if a tear should occur in the bag . . . and as it is just as inert as the Silastic bag, no harm would result."[14] The implant was to be available in three sizes: small (210 cc), medium (270 cc), and a larger size (340 cc) "nicknamed 'The Burlesque' [which] . . . would probably be used only in very large women or certain burlesque queens or strippers desiring an exaggerated augmentation."[15]

The Cronin implant was clearly situated in the history of each of the three procedures—augmentation, reconstruction, and prophylactic mastectomy—of breast plastic surgery. In their introduction, Cronin and Gerow emphasized how the new implant could be used for all indications. (The photographs accompanying the published article, however, show only surgeries for augmentation.)

Explicitly, too, the implant was a response to the criticisms of earlier technologies. Cronin and Gerow listed the pitfalls of earlier technologies: fat transplantation resulting in "fat necrosis and atrophy"; synthetic sponges "that might be pleasing to the eyes, [but] have been characterized by hardness, sometimes rock-like and at all times unnaturally firm." By contrast, with silicone gel implants, "the final result is esthetically pleasing to both visual and contactual inspection."

In their rhetoric, Cronin and Gerow embraced the paradox of the natural—and proclaimed it solved. They argued that women's desire for larger breasts was "natural," the partial result of "the tremendous amount of publicity given to some movie actresses with generous sized breasts." External prostheses were "falsies"—phony, unnatural, something to be endured—but the Cronin implant effected "a pleasing enlargement from within." Designed to be fluid, to mimic "the softness and change of shape with different positions of the body" of the well-developed breast, it provided a "natural feel" as well as look. (It passed "contactual inspection"—one sort of "feel"—and satisfied the women into whose bodies it was placed—another "feel.") Its naturalness was the naturalness of the ideal: "The use of this prosthesis has generally resulted in breasts of pleasing contour, with softness comparable to the young adult breast."[16] Advertisements for the Cronin implant emphasized that it could "approximate the softness, contour and fluid-like mobility of the normal breast." Silicone appeared to be the artificial material whose results were good enough to be claimed as natural.

At the 1964 meeting of the American Society of Plastic and Reconstructive Surgeons, the buzz was over silicone. "Silicones offer great promise as highly versatile, noninflammatory substances for use as tissue implants in plastic and reconstructive surgery," *JAMA* reported in a brief news account of the meeting, "but it is still a promise."[17] Talks at the meeting reflected the same blend of optimism and caution. A presentation by plastic surgeon Ralph Blocksma and Dow Corning's Silas Braley noted the recent explosion of interest in silicone but warned that some restraint was necessary: "It is essential that we pause for perspective in the face of a mounting enthusiasm for the use of silicone implants. We must survey our current knowledge objectively and arrive at a sane, conservative philosophy regarding their use."[18] The most important precautions, they suggested, were to use only medical grade silicone and to take extreme care to sterilize the materials. These warnings implied that the only potential dangers of silicone were those of contamination, thus reinforcing the basic message about the safety of the material itself.

In contrast to the rhetoric of the natural used by Cronin and Gerow and emphasized in advertising, Blocksma and Braley noted the artificial quality of silicone: "It should be remembered that they are, nevertheless, still non-viable materials. . . . they remain forever what might be called, paradoxically, extracorporeal implants, that is, they do not become part of the living tissue."[19]

Despite official precautions, use of silicone was quickly popularized among plastic surgeons in practice. Within a year or two of its introduction, the Cronin breast implant became the technology of choice for simple augmentation,[20] for reconstruction after mastectomy,[21] and as an immediate substitute after prophylactic mastectomy.[22] Surgeons waxed enthusiastic: the new implant "gives the breast an almost normal consistency,"[23] noted one. Another reported that the implant was very soft, caused minimal tissue reaction, was comfortable to the patient and felt natural to the touch, and that the small insertion incision meant less postoperative pain.[24] Most important, the smooth surface of the silicone envelope seemed an effective deterrent to the ingrowth of fibrous tissue that had been the bane of the sponge implant.

Concurrent with the development of silicone gel breast implants came attempts to use liquid silicone as injection material for breast enlargement.[25] Injectable silicone was introduced in occupied Japan in the years immediately following the Second World War. According to a story that may or may not be apocryphal, but is often repeated, material for the first injections came from supplies of silicone fluid used as an industrial coolant and stored on the docks in Japanese cities. Officials noticed that the fluid was disappearing at an alarming rate; investigation revealed that it was being sold to cosmeticians who were injecting it into the breasts of Japanese prostitutes to make them more attractive to American soldiers.[26] Whether or not the particulars of this story are true, it is certain that by 1946 Japanese doctors were using something called the Sakurai formula, a mixture of silicone fluid and unsaturated fatty acids (derived from vegetable oils), to perform wrinkle repair and augmentation mammaplasty by injection.[27]

The fatty acids were necessitated by the propensity of silicone liquid to slip away from the site of injection. The addition of vegetable oil (or other adulterants) had the effect of stimulating production of connective tissue around the injection site, thus providing a barrier to dispersal of the silicone. The theory behind the Sakurai formula bore a striking similarity to the idea that had guided paraffin injection more than fifty years earlier. And, as with paraffin, silicone injection was associated with a long list of readily apparent postinjection local complications, including irritation and inflammation, numbness, tenderness, and discoloration.

In the United States, the histories of silicone implants and silicone injections diverged from the beginning. The most important factor in this divergence was the early involvement of the FDA in regulating silicone injections. Alerted to the potential for the widespread use of silicone injections, and the possibility of serious complications, in 1964 the agency moved to have silicone liquid declared a drug.[28] This classification gave agency officials jurisdiction over the distribution of the material. Thus, beginning in 1965, legitimate use of silicone injections was confined to investigational studies that were monitored by the regulatory procedures of the FDA. These procedures required that investigators apply to the agency for permission to acquire the fluid, that their study designs and results be monitored, and that proof of safety and effectiveness be collected before the FDA would grant approval for general use. Although groups across the country received permission to conduct research on silicone injections, these studies were designed to assess the effects of low-volume silicone injections such as those that might be used to correct scars or to smooth out facial wrinkles. The FDA never approved any studies of injection for the purposes of breast augmentation.[29]

Investigators began reporting the results of their silicone injection experiments as early as 1965. Animal studies looked promising: one group reported "no evidence of systemic effects . . . no significant local reactions other than an early cellular response."[30] But results of other investigations were more disturbing. One study found that silicone injected into mice was taken up by macrophage cells (the immune system's scavengers) and carried throughout the body.[31] And other researchers saw a number of adverse reactions in animals, including nonhuman primates. These reactions ranged from inflammation at the site of injection to atrophy of local fatty tissue to the systemic dispersion of silicone through the lymph nodes, liver, spleen, kidneys, adrenal glands, and blood. Despite this widespread dispersion, the same authors could find no evidence of toxicity and no tumor formation.[32]

Plastic surgeons conducting "legitimate" research on the use of silicone injections condemned their use for breast augmentation. "The first and foremost reason for this attitude," wrote one investigator, "is that the female breast is an organ that supports a high incidence of cancer. Therefore, the blind injections of any foreign material in or around the breast parenchyma is, at this time, a technique associated with known and unknown variables."[33] However, he continued, "the clinical use of silicone liquids in man [sic] preceded any responsible and controlled experiments in animals. These fluids have been used as an injectable prosthesis for at least a dozen years, particularly in Europe and the Orient. Many thousands of persons have been injected and there is no doubt that the clinical use of these fluids has been more widespread than is commonly acknowledged."[34]

In fact, illicit use of silicone injections for breast augmentation was rampant in the United States as well. Contemporaneous accounts in the popular press described one physician who claimed to have injected more than fifteen thousand people. At one point the FDA estimated that in Los Angeles alone there were seventy-five doctors offering silicone injections. In 1967, Dow Corning was indicted for providing silicone fluid to doctors believed to be using it for unapproved purposes—mainly breast augmentation. (The company eventually pleaded no contest to the charge and paid a fine.) Several jurisdictions passed legislation outlawing the procedure.[35] In his 1990 testimony to Congress, Norman Anderson estimated that some fifty thousand women in the United States, mostly in California, Texas, and Las Vegas, had received illicit silicone injections for augmentation mammaplasty in the 1960s and 1970s.[36] By 1971, the AMA felt compelled to publish a warning

> to women who want larger breasts and have turned to liquid silicone injections as the way to have them. "The injection of silicone fluid to increase the size of the female breast is an unapproved surgical technique and is dangerous," said William R. Barclay, MD, assistant executive vice-president for scientific affairs of the AMA. Earlier this summer an unlicensed practitioner in Houston was charged with murder in the death of a Houston woman following injections in her breasts. An autopsy report revealed that the silicone had entered her blood stream and lodged in her lungs. . . . How widespread the practice of breast enlargement by liquid silicone injection has become in the United States is unknown, but it has grown considerably in recent years with the onset of the "topless" fad among waitresses and showgirls.[37]

Such official warnings seemed to have little effect, however. In 1975, *Esquire* magazine published a picaresque account of "Dr Jack," a disgraced physician who made his living from a profitable "itinerant practice" in silicone injection.[38]

At the same time that laboratory investigators were emphasizing the apparent safety of the technique in animals, clinical reports of adverse effects following silicone injection in human beings were building. Such reports first surfaced in Japan in the early 1960s.[39] Similar accounts appeared in Western medical journals beginning about 1964.[40] Typically, these case reports were of young women who had had silicone injections for breast augmentation.[41] Months or years later they would come to a physician complaining of pain and lumps in the breast. Examination would find tissue degeneration and necrosis. Histologic study of the lumps would show that they consisted of silicone, connective tissue, and cells indicative of a chronic inflammatory response. The recommended treatment for such patients was removal of all visible silicone, surgery that often necessitated amputation of the entire breast.

One explanation for such untoward reactions was to blame the lack of skill, or credentials, of the people performing injections—"doctors or . . . skilled but medically unqualified cosmetologists . . . amateurs or . . . 'backstreet' operators"[42]—who were condemned either for faulty technique or for using silicone at all. Most case reports emphasized that the women had their procedures "somewhere else" (Europe or Asia if the author was from the United States, Asia or the United States if the author was European). The implication was that "somewhere else" medical practitioners were unethical and incompetent and that "somewhere else" the controls on unlicensed practitioners were inadequate.

Others blamed the chemical composition of the injection fluid. Most often in these cases, they argued, fault lay with the presence of intentional adulterants (like vegetable oil) or accidental contaminants (as might be found in a dockside barrel) in the silicone fluid. "It is likely that every reported case . . . of silicone granuloma [benign tumors containing silicone] occurring in the human breast has been due to adulterated [silicone]," opined one author.[43] This theory seemed to ignore the fact that nodularities and lumps had also been found in laboratory animals injected with the purest forms of silicone.

Descriptions of women who sought breast augmentation through silicone injection implied that they were young and foolish, at best, or in some cases, that they were immoral by virtue (or lack thereof) of their lifestyle or employment. The *Esquire* piece on Dr. Jack riffed on the "thousands and thousands of sadly self-conscious women out there, women with only a sense of inferiority to stuff into their tank tops and peasant blouses and no *Vogue* subscription to tell them it's okay" and pointed to customer demand as an unstoppable force: "Once the word got out that small-bustedness was not an unredeemable condition, there were just too many women determined to secure, you know, big boobs for a mere government/American Medical Association Alliance to deter." [44]

One author suggested, ever so gently, that qualified and well-meaning plastic surgeons themselves might share some responsibility: "The understandable and justifiable motive of the plastic surgeon to obtain constantly improving functional, cosmetic, and psychological results for his patient has led to the progress of our specialty. . . . Nevertheless, in our eagerness to help, we must be certain to do no harm. . . . In this era of rapid medical publication . . . the responsible surgeon can be pressured or misled into premature use of potentially dangerous materials or techniques."[45]

Throughout the 1970s and into the 1980s, Dow Corning and several groups of plastic surgery investigators continued to research the use of low-volume silicone injections for the correction of facial defects. Officially, these studies proceeded through the regulatory channels of the FDA. Unofficially, it appeared that some investigators were injecting silicone for conditions outside the study protocols and in patients who were never enrolled in the trials.[46] The FDA, however, continued to renew Dow Corning's investigational drug exemption for silicone fluid, allowing the trials to continue. In 1990, as the silicone gel breast implant issue was heating up, the FDA rejected Dow Corning's final report on the investigational use of silicone injections, citing deficiencies in long-term follow-up and a lack of objective outcome evaluation.[47] Since then, despite strong statements from the FDA condemning the use of silicone injections, evidence exists that some doctors in private practice continue to use them.[48]

The lessons learned from the silicone injection experience had little resonance for those pursuing the technology of breast implants. That the dangers of silicone injection might have some relevance to silicone gel implants was discounted by plastic surgeons, largely because the blame accorded operator technique and contamination seemed to exonerate implants.

Warnings about this assumption came only from outside the profession. Chemist Fritz Bischoff refused to distinguish between silicone injections and silicone implants when he wrote a scientific review that "refute[d] the myth that the silicones are biologically inert" and likened the possible outcome of widespread silicone use to that of the paraffin fiasco.[49] Bischoff's objections to silicone, however, were more than scientific.

> Pandering of Madison Avenue to the "little boy" mentality of so many American males must be held accountable for the wholesale tragedy resulting from silicone mammary amplification. . . . While the aftermath—deaths by emboli, necrosis, lumpy breasts and breast amputations after silicone injection—will hopefully sober the exuberance for indulging this form of sexual variance, the genetic outcome of the preoccupation with the female breast may perpetuate a race of top-heavy rather than intelligent women. The effect of surgically implanted silicone and other breasts prostheses—falling in love with a bag of silicone—is in the same category as silicone injections when it comes to mental health. Plastic surgery with organic polymers has reached the dubious state of the art where human beings can be outfitted with a dog-like penis and pig-like breasts.[50]

Evidently *not* a breast man, Bischoff offered a tirade that combined criticism of a consumerist society with misogyny and an evolutionary perspective reminiscent of the French authors who had worried that the ascendancy of bottle-feeding would cause the breasts of future generations to wither. That this warning was publicized by *JAMA*,[51] but ignored by the plastic surgery journals, was probably because of Bischoff's implied criticism of plastic surgeons as catering to a vanity-driven marketplace. As with later critics of breast implants, plastic surgeons could discount such warnings about safety by attributing them to the author's "judgmental" presumption or lack of empathy, a misunderstanding of the pathology of cosmetic imperfection.

In the late 1960s, plastic surgeons began to publish case series that reported the results of several years of use of the silicone implant.[52] While most articles listed hypomastia, atrophy and ptosis following childbearing, congenital asymmetry, replacement of an existing implant, prophylactic mastectomy, and reconstruction after mastectomy for cancer as the indications for surgery, the overwhelming number were performed for augmentation. One surgeon estimated that 70 percent of his cases were for hypoplasia.[53] Cronin himself reported that more than 50 percent of his patients sought surgery for small breasts, while another third came to him to correct sagging after pregnancy.[54] These follow-up reports emphasized the high percentage of good results and made clear that a good result was one that appeared "natural": "To judge the results, one must consider a description of the 'ideal' natural breasts," wrote one surgeon. "Their size should be in correct proportion to the other figure measurements; they should be soft and have a contour with a gentle flow with not too much fullness in the upper half; the nipple should not be lower than the inframammary creases."[55]

The introduction of the Cronin implant, and the early reports of the excellent results achievable in augmentation, also increased the interest of plastic surgeons in prophylactic subcutaneous mastectomy. A 1966 publication described the case of a twenty-three-year-old woman whose doctors had recommended a simple mastectomy for treatment of a benign breast tumor. The patient, while "anxious to be rid of the need for frequent biopsies and the fear of possible cancer . . . was reluctant to lose the breast contour."[56] A subcutaneous mastectomy followed by reconstruction using Silastic implants allowed the patient to retain "symmetrical breasts of good contour."[57] The availability of this procedure assuaged the reluctance of both surgeon and patient, allowing "a common meeting ground where the surgical goal can be achieved without the problems of ablation deformity."[58]

The success of silicone implants prompted plastic surgeons to expand the possible indications for prophylactic subcutaneous mastectomy. In 1970, two surgeons noted that "the development of pure silicone retromammary prostheses has permitted the envisaging—without mental reservation—of an immediate or secondary repair and thus extended to reasonable limits the indications for total mammectomy."[59] Noting that bilateral subcutaneous mastectomy with reconstruction using silicone implants gave "excellent cosmetic results," another surgeon urged his colleagues to increase their attention to prophylactic surgeries for benign breast conditions: "It is sometimes puzzling that benign diseases of other areas, whether or not premalignant, are frequently not treated by limited extirpation of the diseased area, but by resection of the organ system causing the offending lesion [e.g., gallstones]. . . . A concept of treatment based on elimination of the offending organ, therefore, seems pertinent in the treatment of benign breast disease."[60]

Soon, plastic surgeons also took a renewed interest in reconstruction after mastectomy for cancer, an area that long had been mired in the controversy over the natural and the artificial. At the 1970 meeting of the American Society of Plastic and Reconstructive Surgeons, Reuven Snyderman and Randolph Guthrie referred to the controversy as they described their use of the silicone implant for reconstruction. "After struggling for years with rotation flaps and tubed flaps and trying to create the missing mound, with indifferent results, we turned our attention to the use of the Cronin silicone implant."[61] Their procedure was to insert a small-size Cronin implant under the skin of the chest about six months after mastectomy. After the skin had stretched, the original implant was removed and a larger size inserted. Other surgeons reported using the silicone implant in conjunction with abdominal flaps for coverage.[62]

Despite the general enthusiasm, criticisms of the Cronin implant and modifications to its design started to appear as quickly as the implant attained widespread use. Early complaints noted that the implant's shape resulted in too much fullness in the upper portion of the breast, at least until the gel settled. The Dacron backing, designed to address the problem of fixation, proved problematic itself, causing a fibrous tissue reaction associated with a palpable firmness along the edges of the implant.

Published case series noted a number of complications after implant surgery. One report found that of about seven hundred implantations performed by one surgeon over seven years, 33 percent showed a buildup of fluid around the breast, 28 percent experienced a wrinkling of the skin covering the breast, 65 percent experienced firmness, 33 percent hardness, and 65 percent discomfort.[63] Another report described wound disruption and extrusion of the implant, hematoma, erosion of the skin covering the implant, and infection.[64]

One thing the follow-up studies made very apparent was that the supposedly nonreactive silicone implant did cause a localized response in the body. Examination of explanted prostheses revealed that "the body has treated every implant as a foreign body . . . the plastic bag (Cronin) implants have . . . been walled off by a body encystment."[65] At the 1971 meeting of the American Society of Plastic Surgeons, Bromley S. Freeman described the formation of such "body encystment" as part of the process of normal healing, and thus as "inevitable."[66] It was troublesome only when the thickness of the capsule increased, or when the capsule contracted, causing firmness by squeezing the implant. (Freeman was the first to use the term "capsular contracture" to describe this phenomenon.) Surgeons tended to attribute these problems with the capsule to the presence of exogenous contaminants like dust or fingerprints and blamed sloppy operating procedures,[67] or the Dacron backing,[68] while others emphasized problems of surgical technique.[69]

The emphasis on technical causes led to attempts to lessen the complications by changing the implant and the techniques of implantation. Early on, the Cronin implant was redesigned to use small patches of Dacron, rather than the original solid backing.[70]

In 1965, H. G. Arion introduced the first major innovation, the Simaplast, a silicone rubber implant that was filled with saline after implantation into the breast.[71] (The new implant did produce a softer breast but also tended to leak, eventually ending in complete deflation and a reversal of the augmentation.)[72] In 1968, responding to surgeons' complaints, Dow Corning introduced a new and improved Cronin model, distinguished from the original by a thinner silicone rubber envelope, the elimination of the seam around the base, and a teardrop shape.[73] In 1969, at the annual meeting of the American Association of Plastic Surgeons, Franklin Ashley introduced the Ashley "Natural Y," a polyurethane foam–covered silicone implant with an internal Y-shaped structure of silicone rubber. The implant was designed to address the fixation problems of the Cronin—the foam cover provided a medium for tissue adherence—and to allow the shape of the breast to be maintained over time.[74]

Other plastic surgeons described suggested changes in technique. In 1968, W. C. Dempsey and W. D. Latham introduced the subpectoral approach, arguing that inserting the implant under the chest muscle would reduce problems of wrinkling and sagging.[75] Many surgeons, including Cronin himself, reported getting better, more "natural" results by using smaller implants. Surgeons were urged to pay more attention to controlling bleeding and to thorough sterilization procedures and precautionary measures like dipping gloved hands into water to remove all traces of powder. (Surgical gloves are often dusted with powder to facilitate ease of fit.) Perhaps paradoxically, some surgeons shifted from performing the operation in a hospital with the patient under full anesthesia to doing the procedure in an office operating suite using local anesthesia and heavy sedation. These surgeons claimed that their patients reported less postoperative pain and faster recovery.[76] Others noted the economic advantage of performing the operation on an outpatient basis: cost reduction meant more women could afford the surgery.[77]

The in-office augmentation mammaplasty was the subject of an educational film produced in 1972.[78] Funded by a grant from Dow Corning,[79] the film followed Florida surgeon James Baker through such a procedure. As the film begins, Baker is shown sitting at his desk. Jaunty music—reminiscent of the theme from *The Brady Bunch*—plays in the background. Behind him the wall is papered in a patchwork pattern in shades of green. In the forefront, on the corner of the desk and facing outward, is a picture of what one assumes is his lovely family. (Years later, in testimony before the FDA advisory panel, Baker would use his daughter as an example of one of his satisfied implant patients.) Baker describes the advantages of performing augmentation surgery as an office procedure: It is cheaper and easier to schedule; the patient reports less pain and gets an immediate psychological boost.

The other star—the patient—enters. (Baker's nurses play only supporting roles.) She wears a bright red bandanna print minidress, has long, straight, ash-blond hair. In voice-over, Baker outlines the assessment process, explaining that he asks about her health history, especially breast health. He ascertains her reasons for wanting surgery. He sketches the procedure, including the possible complications. They proceed to the photography room for a physical examination and to take the "before" photographs. In one shot, Baker kneels before the young woman, helping her to draw down her surgical gown. He pushes her hair back, then his hands move to her breasts. In a moment, he backs away, takes a picture, motions her to turn to the side.

On the day of the surgery, the patient is shown entering the office, wearing another bright, short dress. In the operating room she is sedated with Valium, washed, draped. (In voice-over, Baker tells us that she will have amnesia for any unpleasantness.) The surgeon makes an incision. The implants are inserted, then he sutures. The woman is helped to sit up. Baker describes the "expressions of delight" as patients look down to see their new breasts. This patient looks pleased but rather dazed and dizzy. After "a soda or a cup of coffee," she is allowed to go home. Baker and a nurse escort her to the parking lot, where she steps into a waiting VW Bug. As the car pulls away, she blows a kiss to the doctor. (In an interview twenty years after the film was made, Baker fondly recalled "this little girl . . . my movie girl.")[80]

Despite the confidence that allowed plastic surgeons to perform augmentation on an outpatient basis, publications during this period imply that the possibility of cancer linked to artificial materials was still a concern. In 1967 and again in 1969, plastic surgeons conducted surveys of their colleagues to try to determine the incidence of breast cancer after implantation.[81] While both studies found a lower than expected number of cancer cases among implanted women, the authors of the earlier study noted the inadequacies of their method and called for more research. The author of the later study, however, seemed confident about his result, speculating that "underdeveloped breasts may be less subject to tumor formation."[82]

By the early 1970s, plastic surgeons seemed assured that cosmetic surgery had achieved some degree of legitimacy within the medical community. "During the last decade the field of aesthetic plastic surgery has emerged from its old dubious position and now has reached the highest level of scientifically-based surgery, performed by skilled and well trained surgeons," boasted one practitioner.[83] The new legitimacy encompassed breast augmentation surgery. One surgeon noted that "the operation is now considered a standard and acceptable procedure by the vast majority of plastic surgeons."[84]

While some located legitimacy in scientific accomplishment, others pointed to the increasing popularity of the silicone breast implants. By 1970, Dow Corning estimated that it had sold fifty thousand implants.[85] Plastic surgeons argued that such sales reflected both the intrinsic demand for the procedure and women's increased awareness of its availability: "In recent years, numerous articles for lay consumption on augmentation mammaplasty have made more and more women with hypoplastic breasts conscious of their small mammary prominences and aware of this type of cosmetic surgery. Consequently it is not surprising that an ever-increasing number of women are consulting plastic surgeons regarding operations designed to correct such defects in body contour."[86] As this quotation suggests, plastic surgeons realized that demand was also the result of increased publicity.

As their popularity grew, breast implants drew other kinds of attention. Although Congress had not yet give the agency authority to regulate medical devices, the FDA's interest in implants was becoming apparent. At a 1970 cosmetic surgery symposium, Dow

Corning's Silas Braley alerted plastic surgeons to that interest: "It is quite likely that the next few years will see the control of mammary augmentation devices by the FDA. In December 1969, a Study Group on Medical Devices was set up . . . and their report has just been issued. It is apparent that there is a widely held belief that it is desirable to have Federal control over implanted devices to ensure their safety and reliability."[87] Braley believed that the government would seek the assistance of plastic surgeons, as expert advisers, in formulating policy toward implants. He urged the plastic surgeons to take their role seriously and to make good use of the opportunity: "*You* will use these materials, and *you* should help to decide what controls will be placed on them. The tighter the controls, the more expensive the acceptable devices and the fewer the innovations. . . . It is up to you as the user, to us in the device industry, and to the Federal Government as controller to work together for the greatest benefit to the patient. Only by such cooperation will the rapid improvements of the past be possible in the future."[88]

Braley's statement pointed out the comfortable relationship between plastic surgeons and industry, and the potential for government regulation to strengthen that alliance. As customers, as sources of technological innovation, and as popularizers of new technology, plastic surgeons had interests largely identical with those of industry. Too much government control could threaten those interests by threatening innovation and profit; these threats against industry would be equally as damaging for the profession. Because the FDA was likely to grant plastic surgeons the status of disinterested expert—while industry, as the object of regulation, could easily fall into an adversarial position with the agency—plastic surgeons could play an important role in determining the status of implants. By reminding plastic surgeons of their shared interests—and common vulnerability—industry hoped to recruit the surgeons to be, if not their proxies, at least their advocates.

Dow Corning was still the dominant manufacturer in the breast implant industry, but it was no longer the only one. In the early 1970s, the number of manufacturers expanded. Most of these manufacturers had ties to Dow Corning. Many still purchased their silicone raw materials from the larger company, and some were ventures started by former Dow Corning employees. Several were owned—in whole or in part—by plastic surgeons. The industry's production facilities ranged from Dow Corning's Midland, Michigan, campus (a model of modern technology) to more ad hoc situations, including at least one company that began operations by producing implants in its owner's garage.[89]

As the number of manufacturers increased, so too did competition to capture market share. More and more variations on the basic silicone implant design became available: patch and no-patch; teardrop, round, or oval; low or moderate profile; seamed or seamless; with or without a foam shell. Implant materials were modified: the silicone envelope was made thinner and the gel looser. Manufacturers introduced a range of second-generation saline-inflatable implants, touting their "superior and more natural tactile properties" and promising that they would "remain softer and more natural to the touch."[90] Another innovation sought to combine the best features of silicone gel–filled and saline-inflatable implants. The so-called gel-inflatable prosthesis was a "closed system" of silicone envelope and a filler tube connected to a reservoir of silicone gel. The tightly folded envelope was inserted through a small incision, then, once in place, pumped full of silicone gel.[91] Another attempt to synthesize the advantages of the two types of implants was the double lumen implant, consisting of an inner envelope filled with silicone gel and an outer chamber that could be filled with varying amounts of saline.[92] One plastic surgeon suggested that the "ultimate breast implant" design would be one that "had a cover which would later dissolve, leaving just the gel in place in a manner that would not leak, drift, or deflate."[93] (Years later, the Dow Corning documents would reveal that the company had tried to develop just such an implant.)[94]

The plethora of technological choices was accompanied by a profusion of procedural choices. A review article published in 1974 gave an idea of the range of these choices by listing the decisions a plastic surgeon planning an augmentation mammaplasty had to make: in-patient or out-patient; incision below the breast or around the nipple; dissection around or through the breast tissue; insertion beneath the chest muscle or just below the breast tissue; implant size, type, and shape; use of steroids and/or antibiotics during surgery; postoperative surgical dressing or brassiere; special exercises or restricted activity after surgery.[95]

Inflatable implants and the softer silicone gel breast implants were essentially shapeless; the ultimate appearance of the breast depended on the surgeon's skill in choosing and inserting the implant and shaping the breast tissue and overlying skin. The aim, as always, was to construct a breast that looked "natural," but plastic surgeons tended to conflate the natural, the normal, and the ideal. One article described the "esthetically pleasing breast" as "one that *looks* balanced. . . . In addition it has a softness to the touch, and it moves with the patient. Normalcy is our objective."[96] A chapter on breast reduction in a plastic surgery textbook was more explicit: "To establish standards of normality, the breasts of 150 healthy volunteers were measured in the standing position with the arms at the side. Of these, 20 were selected as being aesthetically perfect, or nearly so. They were regarded as normal." Because "there were scarcely any differences in the breast dimensions of these women, no matter what their height or weight," the author concluded that "there exists a standard type of breast whose measurements would be aesthetically correct for any woman."[97]

Others disagreed about the consistency of normality, arguing that the "ideal" was always shifting because aesthetics were socially defined. "Lay publications in the 1960s presented small, firm breasts. Throughout the 1970s there has been a progressive change in image to a more ptotic, larger breast as stimulating and desirable,"[98] noted one surgeon. A particular woman's preference would "reflect on the person, her lifestyle, her social level, and the region in which she lives. A prissy, New England school teacher would probably desire smaller breasts than a southern California exotic dancer."[99] Some disputed that the woman's preference should have anything to do with it. "I do not believe that patients are in a position to decide . . . on the appropriate size implant," wrote one surgeon. "They are far too emotionally involved. . . . A surgeon should be expert in aesthetic contour and should be the one best capable of making the proper judgment."[100] Another surgeon, however, pointed out the danger of relying on a (male) plastic surgeon's aesthetic judgment: "To the surgeon [the breast] may remain subconsciously an area of erotic enterprise. . . . A glance through publications on augmentation arouses [sic] the suspicion that surgeons do often build up the unfortunate female's breasts to suit their male concept rather than merely remedy her female deficiency."[101]

One author set a new standard for defining the natural: successful deception. "Except for the incision line," wrote the author, "augmented breasts should be indistinguishable from normal [i.e., nonaugmented] breasts, particularly if the patient were examined by another physician who was unaware that she had the augmentation."[102] (As fans of the *Seinfeld* TV series can attest, the question "Are they real?" remains a powerful cultural theme.)

Fashioning a natural-looking breast proved most difficult in the case of reconstructive breast surgery. For this—and other—reasons, although plastic surgeons had shown increased interest in reconstructive breast surgery, the procedure was not popularized until there was a confluence of changes. In the mid-1970s, the Halsted radical mastectomy was, for the first time, coming under challenge as the treatment of choice for breast cancer in the United States. European studies, the first of which had been conducted as early as the 1940s, had shown no difference in postsurgery survival rates for women who underwent the Halsted procedure and those who had a modified radical mastectomy—an operation which left the chest muscle intact.[103] By 1975, the number of Halsted mastectomies performed in the United States had started to decline. Data from the decade show a trend toward fewer radical mastectomies and more partial mastectomies.[104]

The change in surgical treatment reflected a change in the epistemology of breast cancer. The Halsted radical made sense under a medical conceptualization that saw breast cancer as a disease of local origin and centrifugal spread. By the 1970s, however, the conceptualization was shifting to one that saw breast cancer as a systemic disease in which the breast was simply the first site of occurrence.[105] The shift in conceptualization and, correspondingly, treatment approach opened the way for an increased emphasis on reconstruction after mastectomy.

In 1976, at the annual meeting of the American Association of Plastic Surgeons, Thomas Cronin cited the "trend toward less radical ablations" in noting a recent upsurge of interest in reconstruction.[106] Conservative surgery left more tissue for the plastic surgeon to manipulate. That, combined with recent innovations in implant technology, had improved the results of reconstruction. As results improved, the subject was "publicized in the lay press," and more and more women sought the surgery.

Breast reconstruction aimed to repair the scars and disfigurement left after ablative surgery. Such defects encompassed three problems: the lack of a breast mound, the subclavicular defect (a hollow beneath the collarbone), and the axillary defect (a hollow under the arm). In the 1970s, reconstructive surgery came to include several different procedures: scar revision, coverage of the chest with adequate muscle and skin, nipple-areola reconstruction, and interventions to the other, remaining breast—not just the building of a breast mound, as had been adequate in the past.[107] The choice of surgical technique depended on a number of factors; most important was the amount of skin and muscle coverage available after surgery. Where ample skin and muscle remained, simple implantation of a silicone breast implant might be sufficient. Where skin was lacking but the muscle remained intact, a flap of skin from the abdomen or back might be transferred to the chest and an implant placed beneath it. Where both skin and muscle were scarce, the surgeon might use a rotation flap of muscle and overlying skin from the back or abdomen to cover an implant.[108] Another technique used a temporary inflatable silicone implant placed under the skin of the chest. Over a period of months, the size of the implant was slowly increased, stretching the surrounding skin. When the skin was sufficiently expanded, a permanent implant could be inserted.[109] In the early 1980s, surgeons developed more complex procedures that used the body's own tissues and made an implant optional.[110]

Another change was in expectations for the cosmetic result of reconstruction. Earlier reconstructive efforts were considered successful if they resulted in a breast that was symmetrical with the other breast and provided some degree of cleavage; the gold standard was how the woman looked in a bra or bathing suit. Now, the goal was to construct a breast that looked natural in the nude. The concern with appearance also led surgeons to focus more attention on "the other breast," variously known as the "remaining breast," the "contralateral breast," or the "healthy breast." Increasingly, the other breast was portrayed as a problem—in need of augmentation or reduction surgery in order to match the reconstructed breast in size and shape—or pathologized, conceptualized as particularly vulnerable to cancer and subject to subcutaneous mastectomy with implant insertion both as a prophylaxis against cancer and as a way of achieving cosmetic symmetry.

Shifts in the epistemology and treatment of breast cancer and improvements in the results of reconstructive surgery changed the potential number of women believed to be good candidates for reconstruction. Historically, plastic surgeons had viewed reconstruction as a procedure appropriate only for long-term survivors, agreeing that surgery should be performed only after five or more years without recurrence. (This attitude was a carryover from the Halstedian notion that the success of surgery was to be judged by local recurrence and that reconstruction might cause or mask such a recurrence.) As breast cancer was reinterpreted, however, fears about reconstruction lessened. Debates about timing now focused on how soon after mastectomy one could do the reconstructive procedure and still obtain good cosmetic results. The general consensus was that reconstruction might be performed at about six months postmastectomy,[111] but some advocated that reconstruction be performed at the same time as the mastectomy.[112]

Plastic surgeons emphasized the technical challenges of reconstructive procedures and the satisfaction they gained in performing them successfully. "This is a challenging and satisfying type of surgery because it requires a basic knowledge of breast cancer pathology, an ability to transfer, rearrange, and alter tissue predictably, a strong sense of aesthetic judgment with an eye for symmetry, and a commitment to the volatile emotional needs of the patient."[113] Drawing explicit comparisons between their work and that of oncologic or general surgeons—"while there is basically only one way to remove a breast, there are at least eight different ways to rebuild it"[114]—they crowed about their own achievements: "We can build a breast that looks like a breast, 'feels' like a breast, and 'acts' like a breast in clothing or in the nude (our patients tell us that!). For the first time we can reconstruct a breast from the body's own tissues and without a prosthesis. We can now please the high aesthetic standards and expectations of our patients, their general surgeons, and ourselves."[115]

By the 1980s, technological innovation and improved results had paid off in greater numbers of reconstructive surgeries. The American Society of Plastic and Reconstructive Surgeons estimated that the number of breast reconstructions had almost quintupled over the first half of the decade—from twenty thousand in 1981 to ninety-eight thousand in 1985.[116] As an index to the increased legitimacy of reconstruction, it had became fairly standard for insurance companies to provide coverage for the procedure.[117] The same technological progress, and the same improved results, were available to those performing prophylactic subcutaneous mastectomy. However, as we will see in the next chapter, that procedure had a different fate.

In 1981, a survey of the membership of the American Society of Plastic and Reconstructive Surgeons resulted in an estimate that one million American women had received breast implants since the mid-1960s.[118] Several follow-up studies and historical reviews of implant technology published around this time showed general satisfaction among women who had received the devices and the plastic surgeons who had implanted them.[119] Thomas Cronin, writing in 1978, reflected this satisfaction: "As the era of augmentation is reaching its zenith, the results are more pleasing and the complications fewer." However, he noted, "one major obstacle to perfection remains—the problem of fibrous capsular contracture."[120]

Hardening or firmness of the breast after implantation was, early on, construed as a local complication (a complication estimated to occur in 25 to 70 percent of all implant recipients). At first, as noted earlier, plastic surgeons blamed capsule formation on exogenous contaminants like lint and powder from surgical gloves. Later, however, the problem was reconceptualized as the extension of an essentially normal process. One surgeon attributed hardening to the thickening of the (normally formed) capsule.[121] Others argued that the important distinction was between the "normal" phenomenon of capsule formation and the abnormal one of contracture. Biopsies of the tissue of contracted capsules had found myofibroblasts—"contractile [cells] . . . [which] behave similarly to smooth muscle cells."[122] Capsular contracture, as one writer described it, occurred when the "fibrous capsule contracts inordinately causing an increased hydrostatic pressure within the implant and thus increased firmness. Fibrous capsule contracture is the body's attempt to create the minimum surface area for the contained volume, i.e., a sphere. When contracted, the relatively inelastic capsule tissue surrounding the noncompressible implant contents results in an implant site that feels firm."[123] (In drawing on the laws of physics, the explanation seemed to emphasize that even this was a natural process.)

Contracture was conceptualized both as a reaction to some external factor—an error in surgical technique or postoperative care—and as an anomaly of the individual body—a "potential inherent in certain patients to form this internal scar."[124] External sources under suspicion included hematoma, the Dacron patches still used on some implants, improper surgical dressings, and "chemical and physical irritations (impurities on the surface of the implant, careless management of the tissues)."[125] Problems of surgery—bleeding, infection—could be treated by preoperative avoidance of certain medications and the use of antibiotics. Steroids reduced the inflammatory process. For those patients whose individual susceptibility was the problem, the solution lay in careful "management" of the capsule—a switch to a different implant technology, more care in surgical technique, or use of postoperative manipulation exercises.

Another emerging theory of etiology looked to the phenomenon of "gel bleed." After the introduction of the new silicone gel implant—models with thinner envelopes and looser gels—in the mid-1970s, it was noted that silicone could slowly seep through the silicone rubber envelope. (It had long been known that the silicone envelope could act as a semipermeable membrane for other substances. Early on, this trait had led researchers to suggest that silicone rubber might serve as a kind of timed-release medication delivery system, and some plastic surgeons had taken advantage of the phenomenon by injecting antibiotics or steroids directly into the saline of inflatable implants.) The wide publicity over the effects of liquid silicone in the body (the silicone injection debacle) made the bleeding of silicone a concern. Given the emphasis on capsular contracture and the mystery of its cause, it was perhaps inevitable that the two would be linked. Plastic surgeons had long observed that saline-filled implants seemed less likely to cause contracture. When one study found free silicon (one of the elemental components of silicone) in the tissue of surgically removed contracted capsules, it seemed to confirm the connection.[126]

In 1977, Garry Brody, then chairman of the Medical Devices and Standards committee of the ASPRS, responded to the concerns about gel bleed in an editorial: "It has recently come to the attention of some of the plastic surgery community that the gel-filled implants that are currently available will 'bleed some of their contents through the silicone elastomer membrane. . . . ' this casual observation has led some observers to leap to the conclusion that it is this leak which is responsible for the fibrous contracture. Indeed, a certain paranoia exists that somehow the manufacturers have been attempting to hide this fact from the physician, and that it represents a flaw in their product."[127] Brody reported that he personally had met with the manufacturers, reviewed their science, and was confident that there was nothing to worry about. "While we must continue our search for ways to improve our surgical results," he concluded, "we must not forget that we apparently have a remarkably good product which, while not perfect, appears to be standing the test of time. . . . We must not succumb to pseudoscience, or jump to unsupported conclusions." Although he did concede that "more well-controlled, sophisticated research is required," Brody's implicit message was that the alliance between plastic surgeons and implant manufacturers was still strong.[128]

The intractability of the capsular contracture problem led to the introduction of a whole new array of implant modifications. One innovation was a "low bleed" implant. (A clinical study reported that the implant did not reduce the incidence of capsular contracture.)[129] Several new foam-covered models were marketed, under the theory that the textured surface of the foam would foil the capsule's ability to contract. Instead, plastic surgeons soon discovered that the foam cover tended to tear and disintegrate in the body, leaving a naked silicone shell vulnerable to capsule formation. Later, there were questions about the possible carcinogenicity of the constitutent parts of the disintegrated foam. (In 1990, the concern about cancer caused the most popular model of foam-covered implants to be removed from the market.)[130]

Plastic surgeons continued to modify their surgical technique and postoperative advice, hoping to find the magic formula that would reduce the incidence of capsular contracture. For a while, subpectoral insertion was touted as a solution, then wider surgical undermining to create a larger insertion pocket. Some surgeons swore by strict postoperative immobilization; others by vigorous postoperative breast massage.

Some surgeons explored treatment for contracted capsules. In the standard surgical treatment, called capsulectomy, the surgeon would open the pocket and either remove the capsule or cut around its circumference, thereby releasing the pressure and softening the breast. Other surgeons would remove the capsule and the implant, substituting whichever implant was thought at the time to be the least likely to lead to contracture. Over time, however, the preferred treatment for capsular contracture became the closed capsulotomy. The description of the development of the procedure by its founder, surgeon James Baker, is a testament to the willingness of the plastic surgeon to take his inspiration wherever he could find it.

One of Baker's patients was scheduled for a capsulectomy. The morning of the surgery, she turned up in his office.

> She related to me the following incident:
>
> "I was at a party last night . . . where we were all having a grand time, when a large professional football player suddenly hugged me and squeezed me very tightly. We heard a loud pop and he said 'What was that?' I said, 'I think my beads broke,' and I ran to the bathroom. When I examined myself, my breast had become soft."
>
> Upon examination of the patient the next morning, the breast was indeed soft. . . . I wondered if a surgeon should not be able to reproduce a similar result by manual compression of one of these firm breasts until the described "popping" sound was heard.[131]

(Eventually, as implant rupture became increasingly important in the legal and regulatory wrangling over implants, the practice of closed capsulotomy would emerge as a major fault line in the alliance between plastic surgeons and implant manufacturers. For a long time, however, the only disadvantage noted in plastic surgery journals was that the force necessary to pop the capsule could cause the surgeon to injure the collateral ligament of his thumb.)

In the mid-1980s, the etiological theory of capsular contracture changed yet again. Surgeons began to view the problem as one of contamination of the implantation site by pathogen-ridden breast tissue. "These organisms reside within the breast tissue itself and can be brought into the implantation site at the time of surgery," one article explained.[132] Evidence for the bacterial contamination theory came from two sources: a "highly circumstantial" chain of reasoning linking the known growth of fibrous tissue around bacterial proliferation and the proven presence of bacteria in the breast tissue with the observation of fibrous tissue growth in capsular contracture; and clinical reports of a reduced rate of capsular contracture in cases where the surgeon had irrigated the pocket with an antibiotic solution.[133]

In this conceptualization, the capsule was seen not as a localized reaction to the presence of a foreign body but as a clear-cut immune response to invading microbes. While such a conceptualization allowed the capsule to be seen as a "natural" and protective process, it also provided evidence for the reinterpretation of the phenomenon as the local manifestation of a systemic immune response.

By the late 1980s, plastic surgeons started to reject the idea that capsular contracture was a problem with a technological solution. Rather, individual plastic surgeons began to suggest that it was treatable only by a change of attitude: "Most instances of capsular contracture, even the moderately severe, are more of a personal inconvenience than a surgical disaster. Every experienced surgeon is aware of some patients who have unilateral contracture and complain about the softer side, preferring the firmer breast because of better projection. . . . Patients should be encouraged to accept moderate contractures."[134]

The history of capsular contracture provides scope for several analyses of the relationship of plastic surgeons to silicone breast implant technology. The profession's almost obsessive attention to the problem demonstrates the degree of ownership plastic surgeons felt for the breasts they made. Indeed, the rhetoric surrounding capsular contracture rarely described it as a problem of the lived body. A textbook chapter on augmentation, for example, reported that most women "did not realize that the firm, rock-hard breast was an abnormal and unfavorable result."[135] More often, it was described as a technical conundrum, a barrier to achieving the perfect result. The real target for an attitude adjustment intervention, then, was surgeons themselves.

The development of theories about the etiology of capsular contracture is also revealing. Initial conceptualizations, back during the era of sponge implants, saw the formation of the capsule as a sign of biocompatibility: the growth of tissue showed the body's acceptance of the implant. That the tissue later infiltrated the sponges, causing shrinkage and hardening, was a sign of the inadequacy of the technology. At the introduction of the silicone implant, surgeons courted the capsule, adding Dacron patches as a design element meant to facilitate fixation. As capsules thickened and contracted and the breast hardened, however, the capsule was reconceptualized as a problem, a complication attributable to failures of surgical technique or to individual susceptibility. Later the emphasis shifted to the breast itself as a source of contamination, and the fibrous tissue reaction as a natural, protective response. Lacking in each of these conceptualizations, but to become of great importance, was an explicit analysis of the capsule as the local manifestation of a systemic immune response triggered by the presence of a foreign material.

Attempts by plastic surgeons to solve the problem of capsular contracture by modifying the implant highlights the cyclical pattern through which breast implant technology developed. Initially, all implant technologies were hailed as breakthroughs and were widely disseminated. Then reports of problems began to appear. The initial innovation was modified with changes designed to ameliorate its shortcomings. Eventually all such modifications fell short, however. An entirely new technology was introduced, in the process discrediting the initial innovation. This pattern was repeated with paraffin, with sponges, and with silicone. It continues even today, as plastic surgeons attempt to address recent concerns about silicone by looking for new technological solutions. (In the early 1990s, however, one surgeon acknowledged the finite possibilities for implants, opining that "the perfect breast enlargement or reconstruction will probably not be achieved by implants, but must await further advances in genetic engineering.")[136]

That this pattern was successful is attested to by the improvement in cosmetic results it produced and by the increased popularity of implants that followed. (As the naturalness of the result increased, it became more natural to choose surgery.) What such a technology development process could not do, however, was answer the questions about safety and effectiveness that were the chief concerns of the FDA. Both the structure of the plastic surgery profession and its epistemology proved to be barriers to traditional scientific assessment. (This topic is explored at length in chapter 6.)

In the late 1980s, on the eve of the silicone breast implant controversy, between 100,000 and 200,000 augmentation and reconstruction procedures were taking place each year.[137] An estimated two million American women were carrying implants in their bodies. Surveys conducted by plastic surgeons showed high levels of satisfaction, with 93 percent of implant recipients expressing satisfaction and 96 percent indicating that they would have the operation again without hesitation.[138] The plastic surgeons prepared to meet the FDA with a defensive stance that emphasized the popularity of silicone implants and the responsibility that plastic surgeons had demonstrated as stewards of the technology:

> After 27 years of use, the silicone breast implant continues in popularity with no signs of saturation. It appears to be relatively safe, and despite a variety of outcomes continues to enhance the quality of life for over 2 million women. Thus far its popularity has resisted the concern of governments and the attacks of those who have no empathy for the needs of others. . . . Everyone has the right to control his or her body and mind after duly considering the social pressures and the physical and psychological risks. As plastic surgeons, our responsibility is to provide balanced information and to assess the physical and psychological indications for surgery in our patients.[139]

As the controversy heated up and plastic surgeons faced more challenges to their claims of implant safety, they started to stress the issue of effectiveness, which they conceptualized as the technology's ability to meet women's need for implants. Let us now turn to an examination of how that need was constructed.

4 THE MEDICAL CONSTRUCTION OF NEED (OR, THE "PSYCHOLOGY OF THE FLAT-CHESTED WOMAN")

> Despite the specific sexual significance of the breast, the wish for surgical modification represents much more than a desire for increased sexual attractiveness. . . . the patients seem to be seeking the resolution of various conflicts about self-esteem, affecting the total personality. . . . In a sense, many patients could be characterized as seeking psychotherapy through surgical means.
>
> —Sanford Gifford, "Emotional Attitudes toward Cosmetic Breast Surgery: Loss and Restitution of the 'Ideal Self'"

> But there is another sort of imperfection more subtle by far, and that is the size and shape of the bosom. The woman who feels a need for a change here is involved with the very idea of herself as a woman. This particular goal of perfection strikes deeply into the regions where the turbulent psyche stirs.
>
> —"Bosom Perfection—the New Possibilities," *Vogue*, October 15, 1959

This chapter traces the evolution of ideas about the need for breast implants, with particular attention to how plastic surgeons, together with their psychologist and psychiatrist colleagues, constructed several "diseases" for which implants were the treatment. Such medical constructions of need were linked to the stage of technology development. Did technology development follow need, or did need follow available technology? Were implants the solution for an existing problem, or did plastic surgeons and others work to create a problem for which they had already found a solution? This chapter attempts to answer these questions.

My argument rests on a conceptual distinction between need

and desire. Desire is the individual wish for implants, a web of feelings and knowledges that lead a woman to seek implant surgery.[1] Need, however, is something granted externally. It is recognition and social sanction. Needs are those desires that are legitimized—that is, widely understood to be "genuine" or "necessary" and then institutionalized.[2] In the medical history of breast implants, need was constructed first by pathologizing variation in the size and shape of the breast, making anatomic difference a kind of sickness, and then by promoting implants as the treatment for these newly discovered diseases of mind and body. As the technology improved, and as society changed, the definition of need also shifted.

Need is the product of a complex interplay of individual and society, technology and policy, patient and doctor. Need is a producer, too. It is creative, generative. The medical construction of the need for implants also promoted ideas about the women who desired the devices. These ideas shaped and were shaped by conceptualizations of femininity; the language of need was a gendered one. Changing constructions of need reflected changing conceptualizations of what it meant to be a normal, feminine woman (and, more subtly, what it meant to be a masculine surgeon).

In creating a need for implants, plastic surgeons sought legitimization not only for breast implant technology but for themselves. The epithet "vanity doctor" cut to the quick. To the extent that they could successfully establish implants as the treatment for a number of pathologies (psychological and otherwise), plastic surgeons could disassociate themselves from vanity. (But as we will see, when self-improvement through surgery won more general acceptance,[3] plastic surgeons were happy to redefine the nature and importance of vanity.) Efforts to establish implants as therapy involved promotion to various audiences: other physicians, the lay public. The ways in which plastic surgeons framed their promotional arguments revealed the kinds of relationships that they had with these groups, as well as the relationships they sought.

Finally, need highlights differences between the procedures for which breast implants were used. The need for augmentation was different from the need for reconstruction, and both differed from the need for prophylactic subcutaneous mastectomy. For that reason, the story of need has three parts.

AUGMENTATION: "THE NATURALNESS OF VANITY"

Madame Noel, who practiced in Paris in the 1920s, was one of the few surgeons to devote considerable attention to the issue of why women were seeking cosmetic breast surgery. (Remember that at the time such surgeries were mainly for hypertrophy and prolapse.) She catalogued the most common motivations: a love of beauty; a quest for aesthetic satisfaction; the requirement of certain professions (dancer, gymnast, model) for a certain body shape; *ces cas psychodermatologiques*—women with hypertrophic and prolapsed breasts who had such severe eczema under the breasts that it caused psychological problems; women with oversized breasts, which "when they reach certain proportions prevent those who carry them from leading a normal life, from participating in sports and from dressing normally. These inconveniences frequently cause psychic problems, from simple sadness to neurasthenia to madness and suicide."[4] Noel recommended surgery to women with severe problems but noted that women whose physical problems were less serious tended to be dissatisfied with the results of surgery, in particular, with postoperative shape or scarring. She recollected spending many hours counseling women who came to her in tears over such *anomalies fabriquées*.

In the 1930s and 1940s, plastic surgeons articulated the problem of the breast as one made up of social, cultural, and psychological factors. The power of the female breast lay in its significance as a multilayered symbol. The breast had economic implications: "In the business and social world," noted Max Thorek, "the individual afflicted with abnormal states of the breasts . . . is seriously handicapped, in some instances to the point of desperation."[5] The breast was sexual for the individual: "Woman has always recognized the breast as the focal point. It has always been a major consideration in her mode of dress. . . . she has always been conscious of the power of the beauty of the breast to attract the opposite sex. This consciousness has been a provocative source of libido both to its owner and to those who would possess her."[6] The breast was sexual for society: "During the last decade, magazines, fiction, advertisements, and the stage and screen have combined to accentuate and make a byword of the human factor called sex appeal. . . . the emphasis is on sex and, through the pictures displaying its appeal, we note that it is emphasized by the breasts."[7] The breast was fashionable, highly visible during "sun-worshipping" and in the "scant and scantier attire of all sports activities."[8] The breast was social, a key to "the gates of normal feminine activity" and "social, economic and emotional fulfillment."[9] By encompassing all of these meanings, the breast assumed an almost mystical status: "The human breast may be considered as a physical stigma on the mind or as the mind's focal point upon the physique, and at various times it has been endowed with unusual physical or psychic attributes with its functions, extending into the realms of both the subjective and the objective."[10] (Interestingly, these medical accounts ignore the functional nature of the breast.)

The woman who sought breast surgery was construed as a supplicant, turning to the surgeon for relief of her anguish. "What woman with abnormal breast development," wrote surgeon H. O. Bames, "does not look with envy upon the well formed figure of another woman, and rightly so, for is not a well formed figure the outward manifestation of normal health and development, and thus representative of an ideal to which we all subscribe?"[11] Bames heard yearning in the plaintive voices of his patients—"Doctor, why can I not have a breast which is normal in appearance? I want a breast which makes me look as good as any other woman, if not better"[12]—and he argued that it was the duty of physicians to respond: "As medical men and as counselors, we may not always agree with this view point, but when the problem of such a patient confronts us we must meet it in the best manner possible. While we may never be able to create normal physiology where it probably never existed before, we can and do attempt to restore normal form and contour. By so doing, we relieve any existing physical distress as well as correct the psychic trauma incident to much anomaly."[13]

Although plastic surgeons recognized the importance of the breast, they believed that others did not. They identified a prejudice against cosmetic breast surgery (against all cosmetic procedures, in fact): "The modern attitude toward deformities of the breast is not wholly free of the ascetic and superstitious patterns of the past. Although there is a growing awareness of the economic, sexual and physiologic handicaps imposed by gross abnormalities of mammary form, too often resignation or fear takes the place of constructive efforts for relief."[14] Such prejudice refused to acknowledge the suffering of "the patient whose deformity is not due to trauma or disease, whose life is not endangered by his affliction," and therefore such a patient "is not wholeheartedly conceded the right to seek from surgery the relief to which he believes himself entitled." In contrast, plastic surgeons were able to "hear the appeal of the patient for the right to a place in society, for the removal of a barrier to his economic efficiency, for a share of happiness in his most intimate existence."[15]

Successful surgery could do more than improve the appearance; it could lead to "a change from a feeling of resentment against fate for having inflicted such a stigma; relief from the physical distress and handicap of such a burden; escape from the martyrdom which had to be accepted because there was no remedy; social progress on a par with those endowed with normal physical attributes; in fact a whole new outlook upon life itself."[16] The accomplishment of psychological relief "take[s] this operation from the field of so-called vanity operations in which superficial thinking so blandly puts it, and rightly place[s] it among the most meritorious contributions to human welfare which surgery has made."[17]

While prejudice against vanity accounted for some of the resistance to cosmetic breast surgery, technical and technological inadequacy also hampered its acceptance. As surgeon Jacques Maliniac noted, "Many of the procedures described in the literature are unsubstantiated by adequate clinical experience. Some seem to ignore basic factors in the surgical anatomy of the breast. . . . In certain respects mammaplasty is still largely in the pioneer stage."[18] With the introduction of artificial sponge materials, plastic surgeons claimed to have taken a giant step toward technological sophistication. Their justifications for performing surgery, however, continued to emphasize the personal pain of having less than perfect breasts. Even the disreputable Robert Alan Franklyn pointed to such widespread suffering to justify his practice, estimating that "four million women suffer from micromastia, another ten million from ptosis, at least one out of three of these women suffers psychologically from it."[19]

As sponge implants became the standard of practice, their poor results grew increasingly disturbing to plastic surgeons. One response was to try to ameliorate the problems they caused—hardening, shrinkage—by modifying the technology, but another tactic was to focus on the psychological indications for, and benefits of, augmentation. Although plastic surgeons had always reported psychological indications and results in an anecdotal, casual fashion, the new emphasis on risk called for a more "scientific" and rigorous explication of benefit. (W. John Pangman emphasized this point when he wrote, "The introduction of any non-absorbable foreign substance into the living body involves certain hazards and undesirable effects. When the benefit to the patient outweighs these side effects, the procedure is worthy of consideration.")[20] It was in the 1950s, then, that "hypomastia" changed from being a word used to describe anatomical variation and instead became a diagnostic term, one suggestive of a particular psychological syndrome. Plastic surgeons allied themselves with psychiatrists and psychologists to study the "psychology of the flat-chested woman."[21] This research constructed hypomastia as a "disease" for which implants were the treatment, and aimed to prove the benefits that would balance the risks of augmentation surgery.

The first studies drew upon a psychiatric literature that viewed the breast from a Freudian, psychoanalytic perspective. For example, in 1939 Karl Menninger published an article that focused on the somatic expression of "the unconscious repudiation of femininity in women."[22] The typical somatic correlate of repudiation was "the unattractive angularity characteristic of certain spinsters—the thin, flat-chested narrow hipped neurotic woman."[23] Menninger blamed this "unattractive angularity"on a psychological complex originating in "a frustration in the competition to obtain love from parents." The daughter who was not able to "find some compensation in the female role" learned to "[resent] the more favored and envied males, while secretly striving to emulate them, and at the same time she hates and would fain to deny to her own femaleness. . . . she turns a portion of her hate inward upon her own femininity."[24]

Later psychiatric work also emphasized the connections between breast size, femininity, and psychological dysfunction. An article published in the late 1950s interpreted a preoccupation with breast size as a "serious sign" in female psychiatric patients, one indicative of "a disturbed relationship with a sister . . . a poorly resolved oedipal conflict . . . [an] inability to accept her own femininity . . . [and] poor psychosexual maturation, with an element of narcissism."[25] The breast was significant both as an organ of "feminine identification" and as a body part genitalized to represent a missing phallus. Another article placed even more emphasis on the phallic interpretation, calling the wish for larger breasts "breast envy" and attributing it to "a displacement of shame derived from feelings of genital inadequacy."[26] A woman desiring larger breasts really wanted "to snatch away the male genitals and attach them to herself, in a displaced position, refer[ing] to her wish that she herself could possess male genitalia and thus become a man."[27] Happily, "such women lose the sense of shame and inadequacy about their breasts when they work through their penis envy in analysis."[28]

Thus, one of the early tenets of the psychology of the flat-chested woman was that women who were concerned about the size of their breasts were manifesting a seemingly paradoxical wish to be both more masculine and more feminine: "If she cannot be a man, at least she would be able to be a complete woman with all feminine attributes."[29] Much of the early psychological work on breast implants focused on this conundrum, describing how such women perceived themselves as lacking feminine attributes before surgery—in the process reifying "femininity" by measuring selected constructs—and demonstrating how surgery was able to resolve their gender inadequacy.

The first study on the psychology of women seeking breast augmentation through surgery with synthetic implants was published in 1958.[30] The authors, a psychiatrist and a plastic surgeon, conducted preoperative psychiatric interviews and postoperative psychological and psychiatric assessments of thirty-two women undergoing augmentation with Ivalon sponges. The researchers' main question was "what is the likelihood that [augmentation] will be helpful in treating the emotional problems presented by these patients?"[31] These emotional problems were "latent depressive reactions" expressed "in the complaint of 'inadequate breasts.'" The women described themselves as "empty inside," "hollow," "inadequate," and "unacceptable." According to the researchers, their subjects experienced small breasts as a punishment "related to guilt about affectionate and sexual feelings about their fathers."[32] "It was felt by these patients," the researchers explained, "that all of these qualities could be changed via surgery and the bad feelings would be dispelled."[33]

Surgery seemed an effective way to unravel this emotional snarl. "The results have been unquestionably successful," reported the authors; "not only did the anticipated changes in self-image occur, but the gains have been consolidated with increased social ease, loss of self-consciousness, and curiously, a change in the fixation of feelings on the breast."[34] Their explanation for this success offered a theoretical resolution of the masculine/feminine paradox set forth in psychoanalytic work: "Despite the commonly held view that a woman's breasts are a direct connection with her femininity and therefore with her mother, our data . . . regularly supports [sic] the view that the size of the breasts was felt to be a measure of the father's love. [Patients reported that their fathers were quite passive in comparison to their mothers.] Thus, it might indeed be feminiinty that the girl is getting from her father and that his 'phallic' qualities are more gentle and giving than those of the mothers."[35]

A replication of this dynamic between the plastic surgeon and his patient accounted for the high success rate in augmentation surgery: "This meaning, likewise, could be understood in the unique structuring of the surgeon-patient relationship in which every patient reported the same change in feelings after the operation—i.e., that this made for a miraculous change in inner feelings, as if something she had needed for so long had been given to her (by the surgeon)."[36] The surgeon had "conformed to [the] subconscious fantasies and hopes of the patient,"[37] becoming the nurturing father who gave of himself to make her a feminine woman.

Over the next decade and a half, a series of articles presented at conferences and published in plastic surgery and psychiatry journals drew a profile of women who sought augmentation surgery. Time and again, these women were described as psychologically healthy-appearing individuals whose self-confident exteriors masked their true feelings of inadequacy, low self-esteem, depression, and neurosis. Such women "give the impression of active, competitive 'on-the-go,' people who are physically graceful, often pretty, and socially at ease. They are aware of their attractiveness and seek recognition for it. Evidence of hysterical seductiveness and depressive trends may be blatant or muted. The feelings of anxiety or despair may never be revealed unless the patient is questioned in a psychologically informed manner."[38]

Women often sought implants when they were experiencing a period of turmoil, for example, difficulty in a marriage. For the most part, however, feelings of inadequacy reached back at least as far as adolescence, when "[the girl] noted the breast development of her schoolmates. A sense of inadequacy as a female developed and was aggravated by remarks or jibes of her friends. Fears about her inability to mature physically and a sense of shame appeared, leading to attempts to hide the lack of physical development. Withdrawal from recreational activities ensued and styles of dress became restricted."[39] These feelings could be exacerbated by the media's attention to big-breasted women, but plastic surgeons argued that individual desire was ultimately not a matter of social pressure: "Each woman began by stating that she sought breast augmentation in order to improve her appearance in the eyes of others. . . . When the topic of motivation was pursued, *all* of the women admitted that they would have wished for surgery even if they were 'to live alone on a desert island' for the rest of their lives. It was the private image that counted, not the public one; the surgery was for themselves, not for others."[40] In fact, "good candidates" for surgery were those who recognized the limitations of augmentation surgery. Women who sought surgery to fix an ailing marriage, for example, were to be discouraged. (However, women who wanted surgery to increase their employability as an actress or model or exotic dancer were seen as quite realistic.)

Many of the symptoms associated with hypomastia were sexual. Women reported a reluctance to be seen naked or to have partners touch their breasts. "Those patients who were able to get married reported marital problems resulting from frigidity."[41] While "frigidity" and/or lack of desire on the husband's part were the major symptoms for some women, for others, especially those "whose work is explicitly sexual" (e.g., strippers and topless dancers), "promiscuity" was the prime indication.

Researchers reported that augmentation surgery had positive

results. Women were "happier in all areas of life, with decreased self-consciousness and increased self-esteem."[42] Sexual problems disappeared: "79% reported . . . an increased interest in sex—53% increased frequency, 69% better quality, 52% increased frequency of orgasm."[43] Most patients "reported marked improvement in interpersonal relationships and in many instances the improved attitudes were attested to by the husbands."[44]

Given that plastic surgeons tended to be, by an overwhelming majority, male, it is not surprising that they looked to their patients' husbands for confirmation of the success of surgery. The same sort of gendered view is apparent in the language plastic surgeons used to underscore the trauma of flat-chestedness and the positive impact of augmentation. One surgeon noted that "it is easy to understand how underdeveloped breasts would lead [a woman] to feel inadequate sexually, just as the male feels inadequate with a small or absent penis."[45] A psychiatrist interpreted a woman's "self consciousness" about the position of her newly augmented breasts to her perception of them as being "in a state of permanent erection."[46] In a comprehensive review of breast implant technology and technique, another plastic surgeon wrote that "I advise my patients postoperatively not to look down at their breasts but to look in a mirror. There is an optical illusion of *smallness* on looking down (try looking at your own genitalia and realize how much larger they are if you look across a room in a mirror)."[47] And one physician included renewed "sexual aggressiveness" on the part of the husband on a list of the positive psychological effects of augmentation: "Why? Because the mother of their children, housewife, and confidante is now a new 'chick.' She is sexually desirable again. . . . she has fulfilled his idea of sexuality again."[48]

The natural/artificial paradox that emerged in the development of implant technology was echoed in the psychological construction of the augmentation patient. Plastic surgeons emphasized the inability of "falsies" to improve a woman's psychological outlook. One author explained this phenomenon by describing cosmetic breast surgery as a "restitution of the 'Ideal Self,'" interpreting the desire for implant surgery as "a wish for permanent self-enhancement that would be 'part of themselves' not artificial or added on, but 'real,' like an intrinsic part of themselves that they had always possessed."[49] He noted that the emotional meaning of "real" had nothing to do with the artificiality of the material but meant "'real' in the sense of congruent with the ideal self which has been impaired, even if this 'ideal' existed only in the patient's imagination or at some specific earlier period of her life."[50]

One effect (but perhaps also a cause?) of the focus on the "disease" of hypomastia was the formation of an alliance between plastic surgeons and psychiatrists, two specialties that tended to be ostracized by the larger medical profession. This alliance was based, in part, on shared intellectual interests. As one article noted, "Plastic surgeons and psychiatrists share a joint awareness of the inseparably close relationship which links facial appearance, bodily proportion, and the overall function and structure of human beings to their moral and personal identification, including their sexual fantasies, hopes and fears."[51] The shared interests extended into the financial and the professional. While the psychiatric community profited from the recommendation that all patients seeking plastic surgery routinely undergo an evaluative consultation or a course of psychotherapy,[52] plastic surgeons benefited by establishing among psychiatrists the idea that surgery could be a legitimate treatment option for certain kinds of emotional maladjustment. Psychiatrists helped plastic surgeons in their task of "patient selection," a euphemism that meant choosing patients so as to reduce the risk of malpractice claims. For the psychiatrists, research with plastic surgery patients opened avenues of investigation into a wide range of body image disorders.

Plastic surgeons attempted to promote breast augmentation surgery to other medical professionals as well. In part because speaking directly to the popular press was frowned upon—Franklyn's experience was proof of that—plastic surgeons began using general practitioners and nurses as intermediaries, seeking to establish the idea of implants as therapy among this group with direct patient access. A flurry of articles appearing in general medical and nursing journals stressed the pain of being less than buxom in a "bosom conscious" society, the severity of psychological damage in untreated cases, the materials and methods available to the qualified plastic surgeon, and the excellent cosmetic and psychological results of surgery.[53]

From the mid-1960s into the 1970s, as the silicone gel breast implant was established as the technological standard and the number of augmentation surgeries performed each year increased, plastic surgeons began to take a broader look at the possible causes for hypomastia, expanding their vision beyond individual psychology. Several observers noted that the imperatives of the bosom conscious society, as conveyed through the media, seemed to influence the prevalence of hypomastia. "At present the current emphasis on fashion and form dictates a certain prominence of bosom and many women feel that they are 'generally insufficient' and lack poise if their bust is not adequate," noted one article.[54] By emphasizing the social basis of need, plastic surgeons were able to point to demand as evidence for the procedure's necessity: "Although the surgical correction of the . . . too small bosom to conform with present day ideals set by fashion or current opinion might seem frivolous, the numbers of patients desiring and undergoing such operations shows that there is a necessary place for such surgery."[55]

A more complex dynamic than simple patient demand was suggested in the preface to the 1976 textbook *Plastic and Reconstructive Surgery of the Breast.*

> The number of surgeons proficient in plastic surgery of the breast has grown geometrically. The change in sexual and social mores has made surgery of this area of the body less clandestine than it was before. The popularization of breast augmentation has been partly responsible for encouraging women throughout the world to look at their breasts more critically without feeling too ashamed to seek help. . . . Augmentation mammaplasty, therefore, must be viewed not only as a medical phenomenon but as a technologic and sociologic event. Scientific advances made possible the development and marketing of the silicone prosthesis, but it might not have been discovered and would not have proliferated had not society been receptive and insistent.[56]

Other writers acknowledged the possibility that the increasing numbers of plastic surgeons might have had something to do with the growing demand for breast augmentation: "The past decade has seen a dramatic increase in the number of well-trained, talented plastic surgeons throughout the country. With expert surgery available in every community, public acceptance and demand for plastic and reconstructive procedures has rapidly increased."[57]

During the mid- to late 1970s, studies began to reconceptualize the psychology of augmentation patients. While earlier work had emphasized the pathology of women seeking augmentation, the research published in the latter part of the decade—research clearly influenced both by the women's movement and by the change in mores known as the sexual revolution—focused on the essential normality, and psychological strengths, of women seeking breast augmentation surgery.

Plastic surgeons began to use the language of the feminist critique of health care, particularly the rhetoric of reproductive rights, to combat what they perceived as the lingering prejudice against breast operations. In the feminist literature, the rhetoric of rights was tied to the idea of empowerment, a way for women to stop being passive patients and become active, informed consumers. This consumerist orientation was meant to increase the autonomy of women in encounters with the medical system. As the language was appropriated by plastic surgeons, however, the body itself became a commodity: "Women have become more vocal about their right to normal breasts. Undoubtedly, their political and social liberation have had medical consequences. They are less passive about their bodies."[58] Although the ideology of the women's movement seemed to argue against the demand for breast augmentation, plastic surgeons pointed to the movement as a cause of increased interest in the procedure. Specifically, they noted that the breast was "the only real external evidence of femaleness" remaining in an era of unisex fashion and behavior. [59] In addition, the "no bra" look, "supposedly a manifestation of feminine protest," actually served "to call more attention to an exclusively female symbol."[60]

The trend toward finding psychological normality in women seeking augmentation was exemplified by a study published in 1977.[61] The authors noted that earlier studies "assume and offer evidence that any woman seeking breast augmentation is emotionally unstable." However, they argued, this work was flawed by poor scientific methodology. In their own study, the researchers attempted to correct these flaws by comparing three groups: women who had had breast augmentation, women with average-size breasts, and women with small breasts who had never sought augmentation. Each group was evaluated using a series of standardized tests that assessed general psychological status, as well as measuring self-esteem and attitudes toward things like women, dress, and popularity. Although the authors hypothesized that the augmentation group might show differences in the "areas of self-concept, body image, attitudes toward women's roles in our society, interpersonal adequacy, actual and desired dress, and frequency of sexual intercourse and of social activities,"[62] instead they found that there were few differences among the three groups. In fact, they reported, "the results show the [nonaugmented small-breasted women] to be the most statistically deviant. They scored in more liberated, feminist, assertive, independent and adventurous directions. . . . our mammaplasty women scored as psychologically healthy and . . . comparable to the [average-size women] with whom they seek physical identification. The [augmented] women did differ, however, . . . in their negative identification of their breasts and in their greater emphasis on physical attractiveness and modern or revealing dress."[63] The characterization of the nonaugmented small-breasted women as "statistically deviant" was used as evidence that the desire for augmentation should be viewed as proof of psychological normality. It also suggested that perhaps those women who did *not* seek cosmetic improvement were the ones who were really sick.

As the women's movement problematized the concept of femininity, researchers moved away from emphasizing the ability of augmentation surgery to enhance "femininity" and began to focus on its ability to empower the individual. For many women, one author noted, the result of augmentation was not more femininity but "a more general enhancement of strength and confidence that included masculine and aggressive qualities."[64] The version of self-actualization psychology popular at the time promoted the notion of assertiveness as evidence of a highly evolved personality. Thus this "general enhancement of strength and confidence" signaled that through augmentation an individual could achieve a psychological state superior to mere normal.

Similarly, the 1970s marked a shift in the plastic surgeons' approach to the issue of vanity. Initially, they had denied that vanity played any part in women's desire for surgery (vanity, after all, was antithetical to need): "Vanity is the desire to outdo others, while patients who seek cosmetic surgery are not trying to outdo anyone. They simply want to look normal. . . . The woman with micromastia has a deformity which is difficult to live with."[65] Later, vanity was normalized (naturalized): "When a woman first comes to see the plastic surgeon in consultation she is usually nervous and somewhat embarrassed about the visit. She is ill at ease and wants to apologize for being excessively vain. It is not difficult to reassure her with observations about the naturalness of vanity and the part it plays in all our attitudes."[66] And finally, it was portrayed as a positive virtue, a sign of empowerment: "If all this seems to smack of selfishness, that is exactly how it should be. The primary reason for augmentation should be the honest desire of the patient to feel better about herself."[67]

The depathologization of women seeking breast implants—the journey from need back to desire—vastly extended the potential market for the devices. It also revealed the confidence that plastic surgeons had in silicone implant technology. Just as earlier studies sought to establish a medical benefit to balance the risk of sponge implants, the reconceptualization of their patients as psychologically healthy suggested that plastic surgeons held no serious reservations about the safety of silicone. (Interestingly, in other countries, especially in Scandinavia, plastic surgeons continued to emphasize the pathology of women seeking implants. One explanation for this divergence from their American brethren might be the differing payment structures in these countries: in contrast to the United States, where augmentation surgery was never covered by insurance, many countries in Northern Europe offered coverage under national health insurance programs for women who were found to have psychological indications for the surgery.)[68]

RECONSTRUCTION: "RESTORATION DRAMA"

The European surgeons who developed the first breast reconstruction techniques emphasized the psychological devastation of women following mastectomy. L. Ombrédanne, for example, argued that the mutilation after mastectomy could be more painful than the threat to life posed by breast cancer. Sounding a theme that was to become increasingly important in the development of reconstructive surgery, he described a patient who refused to have surgery for a cancerous tumor unless he could promise to save her breast.[69]

In the United States, it was not until the postwar period, when the demographics of breast cancer began to change, that American plastic surgeons started to focus on reconstruction and thus on the emotional sequelae of mastectomy. Surgeon Harold I. Harris, in an article that described a complicated method for rebuilding a breast, cited the younger population of mastectomy patients and the longer survival time following diagnosis and surgery when he urged plastic surgeons to devote themselves to developing reconstructive procedures. While some women were content to wear a "falsie," he noted, "others find this fate in life unbearable, and their emotional reaction turns them into seriously maladjusted persons. . . . I am of the opinion that the patient is entitled to any and all procedures, whether they be medicinal, physical, psychical [sic], surgical or all combined, in order to give the individual a brighter outlook on life."[70]

Harris's emphasis on the word "entitled" was significant. Throughout the 1950s and early 1960s, reconstruction was controversial. Many plastic surgeons refused to perform the complicated autogenous tissue reconstruction techniques because of the poor long-term prognosis following breast cancer, noting both that "with this shortening of life the surgeon hesitates to inflict a series of operations involving long spells in the hospital" and that the "deforming" scarring at the site of donor tissue could be so severe that "the validity of the procedure is questionable."[71] Sponge implants were feared for their carcinogenic potential. In addition, some practitioners stigmatized patients seeking breast reconstruction "as potentially psychotic and even ungrateful; half-formed suspicions occur that these are weak patients unable to adjust to 'God's Will.'"[72]

But other practitioners were staunch advocates for making reconstructive procedures widely available. The prominent English surgeon Harold Gillies, who pioneered many autogenous tissue reconstructive procedures, advocated immediate reconstruction after mastectomy, noting that the patient would "never know the horror of asymmetry and will be buoyed up [sic] with the presence of a new mammary prominence." According to Gillies, reconstructive surgery had "for its object the removal of the tell-tale scar, a constant reminder of her disaster, and creation of a make-believe substitute which when covered is as good as the 'falsie' and when in neglige [sic] considerably more appealing."[73] The "crying need" that Pangman cited as his impetus for developing Ivalon implants was exemplified by an anecdote about reconstruction: an actress whose "deformity—[she had had a mastectomy]—had caused her to cease work, and become a recluse and a barbiturate addict." After reconstruction with an Ivalon prosthesis, "this patient was off her drugs and again enjoying life."[74] Pangman believed that the availability of reconstruction would prompt women to seek early treatment for breast cancer.

In the midst of the argument, the psychology of women who had undergone mastectomy for breast cancer came under the medical gaze of the psychiatric community. In 1952, *JAMA* published a highly influential study by Richard Renneker and Max Cutler (the latter a surgeon and the former a psychiatrist).[75] Renneker and Cutler described a postmastectomy depressive reaction "marked by anxiety, insomnia, depressive attitudes, occasional ideas of suicide, and feelings of shame and worthlessness."[76] They likened the depression to mourning and attributed it not to "a fear of cancer or death" but to the loss of the breast, "the shocking feeling that the basic feminine role is in danger."[77]

Renneker and Cutler related this reaction to the personal and emotional significance of the breast. They argued that the "two major psychological meanings" of the breast, sexuality and motherhood, represented "the very core of [the] feminine orientation." Thus a woman's reaction to mastectomy would reflect her "deep psychological attitudes of acceptance or denial of the fundamental feminine role."[78] Women who were experiencing a problematic femininity would have the worst reactions: "The trauma tends to be greater in direct proportion to her comparative youth and to the degree of feminine achievement she feels she has not attained (i.e., sexuality, husband and children)."[79]

This construction of the psychological trauma of mastectomy dominated medical thought well into the 1970s. At the beginning of the decade, the process of "mourning the breast" was described as one of the "tasks" of the breast cancer patient.[80] A literature review of the "psychoemotional aspects of mastectomy," published in 1975, concluded that "the predominant psychological reactions to mastectomy are a sense of mutilation and a loss of feelings of femininity."[81] A female psychologist found that "the principal psychological reactions to mastectomy seem to center on the threat to femininity it represents."[82] Although evidence from her own study—"the ultimate drop in self-image is not severe enough to differentiate [the woman who has had a mastectomy] from biopsy or surgical patients"—seemed to suggest otherwise, this psychologist was able to conclude that mastectomy "leave[s] the woman feeling worse about herself and her body" by suggesting that "denial" served "as an apparently successful defense mechanism" for these women.[83] Similarly, researchers who analyzed a 1978 survey of breast cancer patients determined that the "strong indications of successful coping" demonstrated by many women in the survey were simply evidence of their "denial" of the "emotional suffering" wrought by the "sense of mutilation, loss of feelings of femininity, and fear of death."[84]

The earliest work on the psychology of reconstruction, undertaken once the silicone gel implant had become the technology of choice for reconstructive breast surgery, assumed the existence of a postmastectomy psychological syndrome characterized by depression and sadness over the "threat to femininity." "A woman after mastectomy genuinely mourns the loss of her breast, which she feels has removed her femininity," reported one article.[85] Whether this reaction varied depending on patient characteristics—for example, age—was a matter of some dispute. One article asserted that "older patients accept that, at their age, the role of the breast, both functionally and sexually has almost been exhausted. . . . they are prepared to wear an external prosthesis and be grateful for their cancer cure"[86] But another argued that patients who were "postmenopausal and not included in the admittedly 'breast conscious' group" experienced the same reaction.[87]

As with augmentation, the language plastic surgeons used to describe the severity of their patients' psychological trauma reflected their own male perspective: "Indeed, mastectomy in the female has been equated psychologically to amputation of the penis in the male. . . . The patient showing the castration complex after mastectomy may for many months or years have been unable to bring herself to look at the operation area, she may have removed all the mirrors from the bedroom and the bathroom, she may still refuse to let her husband touch her operation area . . . and she may even reveal that her husband has in fact been sexually impotent since the mastectomy, so she requests reconstruction for this very valid reason."[88]

Plastic surgeons set out to show that reconstruction had a wide range of benefits. These benefits included the "structural, psychological, social, and sexual."[89] That is, they argued, women were more physically comfortable after reconstruction—no lopsided posture, no irritating "falsies"; they no longer felt a morbid preoccupation with their illness; they no longer avoided social occasions for fear of being discovered; and they no longer feared that their husbands would be repulsed by their bodies.

Plastic surgeons had confidence in reconstruction but recognized that the procedure had to be promoted so that others could share that confidence. Their promotional strategy included appeals to two groups: surgeons who performed mastectomies and women who had had or might have such surgery. "Education should be directed to both our general surgical colleagues and to the women who face mastectomy so that they all understand that even after the most severe mastectomy the possibility of some type of reconstruction exists," noted one author. [90] However, plastic surgeons had to be on their guard against educating "in a fashion that would become tawdry and eventually evoke a negative response."[91]

The plastic surgeons' attitude toward their surgical colleagues mixed contempt and respect. Plastic surgeons remembered that in the past general surgeons had refused to give them the records of their mastectomy patients and had actively dissuaded women from considering reconstruction.[92] They also accused them of having been indifferent to the suffering of their mastectomy patients, of discouraging discussion of reconstruction "because the surgeon may feel that she should be grateful just to be alive."[93] The guest editor of a 1979 issue of *Clinics in Plastic Surgery* devoted to reconstruction speculated about the reason for general surgeons' reluctance to recommend reconstruction: "Their suspicion . . . is based on the knowledge that breast cancer is a most terrible disease and may not be fully understood by the plastic surgeons who, in their opinion, appear to be most cavalier about the problems and solely interested in reconstruction."[94] Women's reluctance followed the attitudes of their physicians: "The outstanding factor preventing women from having restoration was the fear of recurrence of cancer. This fear stemmed from their general surgeons, who arbitrarily rejected the concept of reconstruction."[95] The quality of the reconstructive result presented another barrier: "The term 'breast reconstruction' is probably an optimistic description of this procedure, since a new 'breast' cannot be created, only a synthetic structure which approximates the original form."[96]

Plastic surgeons knew that in order to establish the legitimacy of breast reconstruction they would have to improve both the results of their work (thus the focus on creating a reconstructed breast that looked natural in the nude) and their relationship with the general surgeons. That relationship was to be nurtured, with territorial boundaries drawn and respected: "Let us refrain from interfering with the judgment of a competent cancer surgeon. . . . Let it be clear that as plastic surgeons we are ready to guide, to help, and to reconstruct, but not to interfere with the survival of the patient."[97] One author suggested a way in which general surgeons might be invited to share in the pleasant empathy between plastic surgeons and their patients: "There is an extra dividend for the ablative surgeon if he accepts the principle of reconstruction. At present he is the *hero* of the cure but ever the *villain* of the mutilation. Once he explains to the patient that she need not go through life mutilated . . . he brings her hope. Many women never bother, but at least they know their surgeon feels it can be done, and in so saying he implies his faith in their cure."[98]

As this passage suggests, some of the benefits of reconstruction were very subtle. Reconstruction was a way to make tangible the optimism of belief in survival (as in the passage quoted above), and it was a strategy for reducing breast cancer mortality: "If reconstruction were accepted universally as the final stage of the treatment program [for breast cancer] . . . perhaps campaigns to achieve earlier detection . . . would be more readily accepted."[99] During the breast implant controversy, both of these arguments would get a lot of attention.

The promotional campaign positioned plastic surgeons as advocates for women and their "right" to reconstruction. They argued that reconstruction should be destigmatized—"the patient's desire for body image restoration is valid and not trivial"[100]—and that all patients should have access to the procedure. In these arguments, plastic surgeons borrowed liberally from the language of the feminist critique of health care: "Since the woman's liberation movement, women have demanded and duly received many more rights and privileges. One of these is the freedom to help decide her own fate when her body is stricken by breast cancer. She has the inherent right to control her own body and her own life. In the past it was a man, a surgeon, frequently a total stranger to her, who made this decision for her [a reference to the practice of women being anesthetized for a biopsy and waking up with a mastectomy if the results showed cancer]. There was no choice. . . . This attitude must be relegated to the archives of the past."[101]

Articles about breast reconstruction in women's magazines quoted plastic surgeons, who used this exposure to promote reconstruction to a lay audience. They argued first that reconstruction was an available possibility, second that its results (both cosmetic and psychological) could be excellent, and third that the resistance of general and oncologic surgeons, as well as much of the general public, meant that women had to be aggressive in seeking the surgery.[102] The "miracle of reconstruction" was, in these articles, embodied by the stories of women who had had reconstruction. Their narratives emphasized that reconstruction was not a "vanity of vanities" but a way to erase "the cancer experience," to become a "complete woman" who was "whole again."

As the results of reconstructive procedures improved, and as reconstruction became more popular, psychological studies reconstructed the women seeking the procedure. While earlier studies had tended to pathologize such patients—"it was believed that a woman who wanted breast reconstruction could not 'adjust to her defect' and was categorically vain, narcissistic, or immature"[103]—the research conducted in the 1980s tended to refute this finding as an unscientific stereotype. Instead, these new studies reported, "women who seek out this operation are noticeably devoid of any clinically significant psychopathology, are usually well-adjusted, and maintain a high level of psychosocial functioning."[104] The desire for reconstruction came to be portrayed as a positive strength: "The desire to want to change an undesirable condition that is indeed alterable may be evidence of adaptive responses and flexible coping."[105] As one author interpreted it, the wish for reconstruction represented "a state of rebellion against [passive] 'acceptance.'"[106] And the struggle against acceptance was a struggle toward healing. (By implication, then, pathology now resided with those women who did not choose reconstruction. One study, for example, looking at the question of why women did *not* seek reconstruction, found that they tended to be motivated by fears of recurrence, by a martyr complex, by a feeling that the mastectomy site should remain as a badge of honor or rage, by a distorted perception of the likely results, or by an attachment to the external prosthesis.)[107] The psychological met the political in the language of choice: "Women are beginning to demand a role in crucial decisions concerning their own bodies. In the struggle for self-acceptance, the process of active choice sustains one's sense of personal competence."[108]

PROPHYLACTIC SUBCUTANEOUS MASTECTOMY: "SOME QUIXOTIC USES OF INVALID DATA"

From the beginning, plastic surgeons described prophylactic subcutaneous mastectomy as a preventive procedure that, given the availability of implants (or other methods) for reconstruction, would be acceptable to women. In the first report about prophylactic subcutaneous mastectomy, published in 1917, Willard Bartlett made a statement that was to resonate through the history of prophylactic mastectomy. "The breast is of such psychic importance to the female patient," he wrote, "that it is usually late in the course of breast tumors that the surgeon is consulted. It is the fear of having the breast mutilated that keeps patients away and allows a tumor to run a progressive course."[109] Bartlett argued that because what mattered to women was the form of the breast, not the function, an operation that could remove vulnerable breast tissue while maintaining the overall shape of the breast would be both medically effective and psychologically acceptable and thus would serve the objective of prevention.

Bartlett's list of indications for the procedure included fibroadenoma and fibrocystic mastitis, conditions he believed to be pre-cancerous. When, in 1951, the subcutaneous mastectomy was once again taken up in the plastic surgery literature, it was for an indication described as "multiple cystic disease of the breast."[110] To the authors, indeed to the medical community as a whole, such conditions were problematic: they were in themselves a source of pain and suffering to the patient, and they were also thought to be premalignant. The treatment dilemma they posed was one of aggressiveness: "simple" mastectomy (complete removal of the breast) was thought to be "too radical," but doing nothing at all seemed too great a risk.

The outlines of the subsequent history of prophylactic subcutaneous mastectomy are apparent in these early reports. That history has three important strands: the belief that certain "benign" conditions (like "fibrocystic disease") were premalignant; the argument over how much tissue had to be removed for the surgery to be truly prophylactic; and the evolution of improved methods for reconstruction after mastectomy. The idea of prophylactic mastectomy rested on two very important assumptions: that it was the fear of mutilation that prevented women from seeking treatment for such conditions before they turned deadly, and that the substitution of an artificial breast could be not only acceptable but preferable to living with uncertainty.

As with augmentation and reconstruction, Pangman promoted his synthetic sponge implant design for use in prophylactic subcutaneous mastectomy. He wrote that this application "represented the greatest opportunity for good," explaining that "carcinoma can certainly be prevented and women who resisted ordinary mastectomy or even biopsy will often welcome this new operation." In fact, he asserted, subcutaneous mastectomy followed by reconstruction with Ivalon implants resulted in a "post-operative appearance and comfort [that] is without exception an improvement over the pre-operative."[111]

Despite Pangman's optimism, the adverse reactions seen with sponge implants proved to be most severe in women who had had subcutaneous mastectomy. Hardening, infection, and extrusion were more acute because of the lack of tissue between the implant and the skin of the breast.[112] Introduction of the Cronin implant, designed to minimize these poor results, seemed to revitalize the performance of the procedure.

Prophylactic subcutaneous mastectomy followed by reconstruction with a silicone implant was touted by many plastic surgeons, including Cronin himself, but differences in terminology and—as with all uses of breast implants—no standard system of reporting make it difficult to know how many prophylactic procedures were being performed, and for which indications. (And, especially, with what result.) A survey conducted in the mid-1960s found that almost half of the responding plastic surgeons had performed the operation.[113] A 1969 survey, however, seemed to indicate that only a small proportion of implantations using breast prostheses followed prophylactic procedures.[114]

The seeming reticence of plastic surgeons may have been a result of the conflict over the nature of prophylaxis that was brewing in the literature. There were two main camps among plastic surgeons. Those in the first camp believed that the operation should be performed as a palliative to the immediate suffering attendant with benign breast conditions—suffering which included repeated biopsies and fear of cancer—but they were uncertain of the procedure's ultimate effectiveness in cancer prevention. Those in the second camp believed that the procedure, properly performed, could prevent the development of a malignancy. Most plastic surgeons fell into the former camp: given the reports of cancer found after subcutaneous mastectomy, [115] the continuing concern with the possible carcinogenicity of silicone implants, and the lack of any real follow-up of prophylactic mastectomy patients, cancer prophylaxis was a problematic notion.

Ambivalence over the effectiveness of prevention led to a debate over indications and technique. Recommendations for surgery encompassed conditions ranging from "benign tumors, cysts, [chronic cystic mastitis], and mastodynia [breast pain]" to "cancerphobia."[116] Despite attempts to classify the histology of "fibrocystic disease" and correlate it with future malignancy, there were no clear, authoritative indications. In practice, most plastic surgeons seemed to perform the procedure for patients who presented with lumpy breasts, pain, a history of biopsies or cancer scares, perhaps a family history of breast cancer, fear of cancer, and—most important—a desire to keep the shape of their breasts.

The main promise of prophylactic subcutaneous mastectomy was to offer some relief from worry while maintaining the breast form. To keep this promise, plastic surgeons had to perform a delicate balancing act. Removal of too little breast tissue, although improving cosmetic results, made a mockery of prophylaxis. Removal of more breast tissue, however, meant distinctly poorer cosmetic results and thus—for both surgeons and their patients—a lack of enthusiasm for the operation. This tension was reflected in two publications that appeared in the early 1970s.

In 1973, Dow Corning published a monograph on the indications for and surgical techniques of prophylactic subcutaneous mastectomy.[117] In a letter to the editor of *Plastic and Reconstructive Surgery*, plastic surgeon Reuven Snyderman objected to the publication. Citing a passage that read "one must . . . produce a skin flap covering that is adequate to retain and cover the prosthesis. Uniformity of thickness is also required to obtain a good cosmetic result,"[118] Snyderman accused the authors of having advocated fitting "the operation to the patient." "If subcutaneous mastectomy is to be of most value, so we can honestly look our postoperative patient in the eye," Snyderman argued, "we must make a strenuous effort to remove as much of her breast tissue as possible—without being overly concerned about the appearance or the viability of our remaining flaps."[119] "We, as plastic surgeons, have become deeply involved in breast problems—and rightly so," Snyderman concluded. "If this is to remain rightly so, however, we must not be in the position of trying to sell an operation to our colleagues and the general public. An operation should never go searching for patients."[120]

Prophylactic subcutaneous mastectomy put the specialty in a precarious position. As long as the procedure lacked general legitimacy in the medical profession (legitimacy that would not come until its effectiveness was demonstrated), plastic surgeons would have a monopoly on performing it. But the same lack of legitimacy made it necessary for them to promote the procedure. Part of that promotion entailed promising more than a curative procedure—promising, in fact, an aesthetic result that they knew could be achieved only by gambling with the prophylactic intent. As the debate over prophylactic subcutaneous mastectomy continued, a deep schism developed within the profession, with one side arguing against performing the procedure (in part on the grounds that it was harming the reputation of the specialty) and the other side determined to prove the value of the operation "scientifically."

At the 1974 annual meeting of the American Society of Plastic Surgeons, Erle Peacock issued a critique of subcutaneous mastectomy that pointed out the generally inadequate understanding of its proper indications and, thus, its wide misuse. Peacock argued that breast lumps and nodularies were of hormonal etiology and that "fibrocystic disease" was not a precursor to cancer. The propensity of plastic surgeons to recommend subcutaneous mastectomy for all breast lumps—(Peacock gave the example of a recent question on the board certification examination set by the American Board of Plastic Surgery)—"strongly suggests that benign disease of the breast is not thoroughly understood, and that the utilization of Latin names and dependence upon a pathologist's diagnosis may be responsible for some expensive and mutilating surgery which could have been averted."[121] The argument that plastic surgeons were using phony scientism to justify their use of prophylactic subcutaneous mastectomy was devastating enough; his suggestion that they were motivated by profit and the description of the result of such surgery as "mutilation" approached heresy.

The counterargument was presented the same year, at the annual meeting of the American Society of Plastic and Reconstructive Surgeons, where Vincent R. Pennisi—the man who was to emerge as leader of the movement to prove the effectiveness of prophylactic subcutaneous mastectomy—made his own statement. He began by explaining the logic of his focus on prophylaxis: "The mortality [from breast cancer] has increased, in spite of early detection, education, modern screening techniques, and the best surgical and radiological therapy available. . . . Early recognition and surgery have improved upon the 5-year and 10-year 'cure' rates, but we must still contend with the awesome reality that one in 15 women will at some time acquire breast cancer. Interestingly, one in 10 women has gross fibrocystic disease. . . . We believe more attention should be directed toward one of the precursors of breast cancer, namely fibrocystic disease."[122]

To bolster this position, Pennisi presented data that he and a colleague had recently collected (with the financial support of Heyer-Schulte, an implant manufacturer). Pennisi surveyed all board certified plastic surgeons in the United States, asking how many prophylactic subcutaneous mastectomies they had performed and how many times occult cancers had been identified in the tissue removed. The 460 plastic surgeons who responded reported performing 4,179 subcutaneous mastectomies in the past ten to fifteen years. In 82 of these cases, pathological examination of the tissue found occult carcinoma (cancers that had not yet been detected). "In a further breakdown of the statistics," Pennisi continued, "it was found that the 82 cases having occult carcinomas were operated on by 53 surgeons who reported 1385 subcutaneous mastectomies . . . an incidence of 6%." Significantly, he added, "this 6 percent incidence is close to the incidence of breast carcinoma in the female population." The same data showed that only twenty-four patients developed cancer after the subcutaneous mastectomy, "an incidence of 0.5 percent." "At present," he concluded, "prophylactic removal of all the breast tissue appears to be the only effective way of preventing breast cancer in the obviously vulnerable woman."[123] Pennisi's apparent point was that women with fibrocystic disease who had had subcutaneous mastectomy were less likely to develop cancer than women who had not had the procedure, but his lack of methodological sophistication (self selection in responding to a survey; further selection of which data to analyze; no standard indications for surgery; no control group) rendered his analysis essentially useless.

The 1974 survey formed the basis of the Subcutaneous Mastectomy Data Evaluation Center (SMDEC), a national registry directed by Pennisi, which over the next decade collected data from plastic surgeons about their prophylactic mastectomy patients and issued periodic statistical analyses of these data that purported to "prove" the effectiveness of prophylactic surgery.

Pennisi stressed the medical (as opposed to cosmetic) nature of the procedure and pointed out the need for teamwork by patient, pathologist, and plastic surgeon (a triangulation that echoed the relationship among patient, plastic surgeon, and general or oncologic surgeon in the area of reconstruction).

> The cooperation of the patient, pathologist, and surgeon is of utmost importance. The patient must be properly motivated and accurately informed. The pathologist must describe the biopsy specimen in detail and specify not only the varying histologic types of fibrocystic disease present but also the degree of severity. . . . The surgeon must be sufficiently knowledgeable regarding the prophylactic approach to breast cancer and fibrocystic disease so he can properly inform his patients of the pros and cons of the operation. . . . It must be emphasized that this is *not a cosmetic operation* but one which is aimed at reducing the incidence and mortality of breast cancer in a woman who may be at greater risk because of the presence of moderate to severe fibrocystic disease.[124]

But other voices urged caution. In a 1977 editorial in the *New England Journal of Medicine,* Harvard plastic surgeon Robert Goldwyn argued that although the "popularity [of prophylactic subcutaneous mastectomy] is increasing, this procedure has not yet received enough dispassionate scrutiny."[125] The indications were still "controversial and imprecise," he wrote. The "inability of the surgeon to remove with certainty all breast tissue" continued to be a major barrier to ensuring prophylaxis. Goldwyn urged his colleagues to contribute data to the SMDEC but warned "it may be many years before subcutaneous mastectomy can be assessed with objectivity. Until then, this operation should be neither blindly extolled nor summarily rejected."[126] Other authors echoed this kind of concern, noting the existence of two classes of indications for the operation: "1) indications related to cancer, either as treatment or prophylaxis, and 2) indications not related to cancer." [127] While the former were problematic, as the authors noted, because the "existence of 'premalignant' breast lesions is not universally accepted," surgery for non-cancer-related indications also presented dilemmas: "The patient must be aware that she may be trading relief of pain for deformity."[128] Given the lack of scientific support for the operation, it was troubling that "the continued enthusiasm among plastic surgeons . . . may lead to an increased demand on the part of both general surgeons and patients."[129]

While these objections focused on the fear that prophylactic subcutaneous mastectomy was too radical a procedure for the reasons it was being performed, other opponents of the procedure objected that it was not radical enough. As Reuven Synderman wrote, "I think it would be advisable to eliminate the term subcutaneous mastectomy and state clearly that when surgical intervention is warranted it would be much better to perform a mastectomy. . . . If the physician is to put this woman's mind at ease as well as to believe that as much as possible is being done, then the subcutaneous mastectomy is not the operation of choice."[130]

In 1980, Pennisi issued another report on the work of the SMDEC: in one thousand patients followed for between one and nine years, there had been two cases of breast cancer. He blamed these failures of prophylaxis on the poor surgical technique of the surgeons who had performed the operations, and refused to consider the suggestion that the procedure itself might be ineffective. Perhaps in response to the critics who had noted the dilemma of "trading relief . . . for deformity," Pennisi once again emphasized the noncosmetic nature of the operation, asserting that patients selected for surgery recognized and accepted this: "Their breast disease and vulnerability to cancer far outweigh their concern for aesthetic breasts. After all, how many women in the 40–year age group have truly aesthetic breasts? However, their concern for the total loss of one or both of their breasts can be great. Therefore, they willingly subscribe to the prophylactic treatment of breast cancer and readily submit to a subcutaneous mastectomy to salvage their breasts and their lives."[131]

Pennisi also took his case to the popular press. In a 1981 article in *Good Housekeeping* that reported the controversy over prophylactic subcutaneous mastectomy, Pennisi was quoted as saying, "It all boils down to needless surgery versus needless loss of life. If I do ten prophylactic mastectomies and nine are needless, it's still worth it to save one life. The nine at least gain peace of mind."[132] The seeming callousness of this statement, a callousness that was also apparent in surgeons who recommended that the patient be made to wait for several months after subcutaneous mastectomy for implant insertion in order to "allow [her] to see her deformity and not to expect that her breasts can miraculously be made as before,"[133] was unlike the usual empathy of the plastic surgeon and seemed to highlight the curative nature of this procedure.

As the controversy over prophylactic subcutaneous mastectomy continued, it became increasingly passionate and often personal. In 1983, Loren Humphrey, a surgical oncologist, published a scathing critique of the procedure. Subcutaneous mastectomy was being performed on women for such a wide range of indications that the idea of prophylaxis had no meaning, he argued. Rather than addressing the questions of which conditions were real risk factors for breast cancer and whether subcutaneous mastectomy was truly prophylactic, proponents of the procedure had based their support on "opinions and some quixotic uses of invalid data."[134] He singled out "Penissi" [sic] for much criticism, noting that other—anecdotal—attempts to determine the incidence of cancer after subcutaneous mastectomy had suggested rates much higher than those found in the SMDEC data. For example, in a sample of sixteen women who had undergone subcutaneous mastectomy with prophylactic intent, Humphrey himself had determined that three of them later developed breast carcinoma. Joining other critics, Humphrey argued that any condition serious enough to be a precursor to breast cancer was better off being treated with a total mastectomy.

The following year, an army plastic surgeon stationed at Walter Reed Hospital raised another issue when he published a follow-up study of eighty-eight patients who had undergone prophylactic subcutaneous mastectomy between 1965 and 1979. The results showed a 25 to 30 percent incidence of acute complications following insertion of a silicone implant, not including a 100 percent occurrence rate of capsular contracture. Concluding that prophylactic subcutaneous mastectomy was linked to high rates of morbidity, the plastic surgery department at Walter Reed had stopped performing the procedure, opting instead for total mastectomy, but only in "women who are severely symptomatic or who by reason of histology or family history appear to be at significant risk."[135]

In a comment appended to the report, plastic surgeon Jack Fisher was also highly critical of the procedure. Calling evidence for the effectiveness of prophylaxis "shaky," Fisher argued that plastic surgeons' insistence on pursuing prophylactic subcutaneous mastectomy was serving only to isolate them from other medical specialties and hinder the growth of the profession in other areas: "Most plastic surgeons wish that they could gain the interest of their general surgical colleagues in more routine breast reconstruction following mastectomy. Perhaps this alliance will be possible only after we shed our persistent advocacy of subcutaneous mastectomy for cancer prophylaxis. If I am wrong and biologic justification can be established for [subcutaneous] mastectomy, why then have we been unable to find that evidence and express it in terms acceptable to our biostatistical, oncologic, and epidemiologic colleagues? They want to prevent breast cancer as much as we do."[136]

Pennisi responded to these critics in a letter published in *Plastic and Reconstructive Surgery*. Noting a number of errors in Humphrey's article (including the misspelling of Pennisi's own name), he argued that such carelessness was emblematic of the author's total misunderstanding. He rebutted the significance of the incidence of cancer in Humphrey's sample of sixteen women, arguing that "I expect that the remaining 13 patients will also develop carcinoma because of inadequate removal of breast tissue. This is not a failure of the operation but merely a failure of the surgeon."[137] He reiterated his argument that prevention through surgery would be the only effective cure for breast cancer. Finally, he blasted Fisher for his observations.

> Dr. Fisher suggests that we as plastic surgeons should make a deal with the general surgeons. We should cease doing prophylactic mastectomies and instead permit the general surgeons to amputate the breasts. They in turn will provide us with total breast reconstruction. Where does the breast cancer patient fit into this deal, or doesn't anyone care. I am appalled by the thought that some of our hallowed halls of medical learning are imparting this kind of insensitive thinking to our residents in plastic surgery. Such misguided teaching can lead to a clouded surgical judgment and a compromising of ethics, and from my vantage point there appears to be a lot of that going around![138]

Pennisi soon buttressed his conviction with a report on the most recent analyses of the SMDEC data. The center had collected information from 165 surgeons on 1,244 patients with an average of seven years of follow-up. At the time of surgery, occult breast cancers had been found in 4 to 5 percent of these patients. In the remaining group, only six women—0.5 percent—had developed cancer. "These data indicate," Pennisi wrote, "that, in fact, the subcutaneous mastectomy is an effective prophylactic operation against breast cancer. Certainly this result should quiet those critics who advocate total mastectomy or simple mastectomy because the subcutaneous mastectomy does not remove all of the potentially vulnerable tissue."[139]

One critic who remained skeptical was Loren Humphrey, who, in a comment appended to Pennisi's report, suggested that readers consider several questions as they looked at Pennisi's data. First, which histopathological changes associated with benign breast disease placed a woman at greater risk of cancer? Second, what was the expected incidence of cancer in women with benign breast disease? Third, what was the expected incidence in women without benign breast disease? As Humphrey well knew, Pennisi's work addressed none of these questions. Given the murkiness of surgical indications and the lack of controls, his numbers could prove nothing.

Humphrey went on to dispute the very basis of the concern with "fibrocystic disease" as a premalignant condition. "Most women with pain, tenderness, and lumpy breasts do not have disease," he asserted, "but only a reflection of hormonal changes during the menstrual cycle."[140] In fact, this view seems to have been representative of the larger medical community. A contemporaneous article in the *New England Journal of Medicine* (Susan Love, later to achieve great fame as the author of a popular book on breast cancer, was the lead author) reviewed the literature on fibrocystic disease and concluded that the diagnostic category "fibrocystic disease" was conceptually useless as a risk factor for breast cancer.[141] (Interestingly, Love and her colleagues virtually ignored the large plastic surgery literature on the subject of fibrocystic disease and prophylactic mastectomy. Pennisi's name, in particular, was conspicuous in its absence.)

Whether driven by the controversy over prophylaxis or the poor clinical results, the statistics on the numbers of subcutaneous mastectomies performed during the 1980s show a sharp drop.[142] National Center for Health Statistics data from hospital discharge summaries report fourteen thousand procedures in 1983; in 1987, the number was five thousand. In subsequent years it became so small that it was no longer reported.

In 1987, at the Ninth International Congress of the International Society of Aesthetic Plastic Surgery, Pennisi presented a "final statistical analysis" of the data gathered by the SMDEC since 1975. Pennisi's presentation was, to put it kindly, confused. The center had accumulated data on 1,500 patients, he reported. About 70 percent of these patients had been followed for an average of nine years, while 30 percent had been lost to follow-up entirely. Six patients had "acquired" breast cancer after subcutaneous mastectomy. Pennisi represented this as an incidence of less than 0.5 percent in the 70 percent of patients retained in the study and a "projected incidence" of 0.57 percent in the entire sample. (In fact, these numbers should be reversed.) A table of results on the 1,500 women in the original sample categorizes the results as: 67 percent cancer free; 1 percent not cancer free; 30 percent lost to follow-up; 1 percent expired from breast cancer; and 0.3 percent expired from other causes.

Pennisi elaborated on his idea of prophylaxis by comparing breast cancer to lung cancer, noting that "both of these fatal diseases can be prevented by prophylactic measures."[143] "Altered smoking habits have significantly affected lung cancer and heart disease in males so we know prophylaxis works," he argued.[144] Perhaps drawing on the rhetoric being used to promote augmentation and reconstruction, he framed the issue of prophylactic mastectomy as one of patient rights: "A patient who has had multiple biopsies, experiences an uneasiness that becomes intolerable and that patient should be entitled to the option of prophylactic therapy."[145] And, as with the two "cosmetic" procedures, the ultimate good result was the "many unsolicited comments by patients who had the procedure and were grateful to their doctors who performed it."[146] "The purpose of this project," Pennisi wrote, "was to prove that a thoroughly performed subcutaneous mastectomy would prevent breast cancer in a high-risk woman. In addition, it was our hope to attract interest in the prophylaxis of breast cancer rather than continuing down the futile path of breast cancer therapy. Both of these objectives have been met."[147] "It is encouraging to see," he added, "that our concepts were correct though premature."[148]

It was the technology of silicone gel breast implants that allowed prophylactic subcutaneous mastectomy to be promoted as an option. However, in contrast to reconstruction, the procedure was never fully legitimized by the larger medical community (or even among plastic surgeons). Despite Pennisi's efforts, plastic surgeons who performed the procedure were never viewed as practicing a serious form of preventive medicine. Well into the 1990s, plastic surgeons continued to debate the merits of prophylactic subcutaneous mastectomy, but the topic was largely elided, unacknowledged, during the public performances of the controversy over silicone gel breast implants. Although plaintiffs in two of the highly publicized cases against the manufacturers had had the procedure, as had many of the most vocal activists (on both sides), the public issue was framed, perhaps for simplicity's sake, as one involving augmentation and reconstruction.

Since the discovery of the BRCA genes—genes linked to a particularly virulent form of breast cancer—prophylactic mastectomy has once again become a topic of discussion and debate. (There has not been a focus on subcutaneous mastectomy, however.) The media have presented many stories of women "faced with a terrible choice." The choice, of course, is whether to have their breasts removed after they learn that they carry the breast cancer gene; the phrase used is that these women are "living with a time bomb." Implicit in these stories is that women consider this option only because of the availability of reconstructive surgery. The extent to which prophylactic mastectomy provides true prophylaxis is still uncertain,[149] yet these narratives of the dilemma are little concerned with the actual protection afforded by the surgery. Rather, the emphasis seems to be on control: will you live with the "time bomb," or will you take action?

At the beginning of this chapter I hinted at some of the subtexts beneath the medical construction of need. Among them was the gendered nature of need. Psychological research on women seeking augmentation or reconstruction began with the assumption that they were emotionally unstable. Given the power of the breast as a symbol of femininity, this emotional instability was conceptualized as a problem of deficient femininity. When researchers set out to assess the effectiveness of breast surgery, they first had to quantify the extent of this deficiency and describe its parameters. That is, the measurement of deficient femininity required that femininity itself be conceptualized as a measurable entity and operationalized using specific variables. In the process of conducting "scientific" research on women, femininity necessarily was reified.

The areas of investigation selected reveal that reification. The early interview-based studies focused on the woman's roles of wife and mother, on her social functioning as represented by participation in dating and heterosexual relationships, on appearance and the search for stylish clothing, on a subjectively defined "happiness," and on the importance of body image to general self-esteem. The femininity portrayed in these studies constructs women as creatures existing in relation to others—first parents, then husbands and children—who demonstrate appropriate, but not obsessive, concern with appearance. Proper fulfillment of the feminine role requires a woman to be (and to look) outgoing, social, and happy. Her self-esteem is inextricably linked to the performance of that role. Implants' curative power lay in their ability to remedy any deficiencies in this femininity. "Happiness" increased as women were able to wear new styles or sizes of clothing. Body image and self-esteem improved and led to better role performance: single women dated more; married women had more, and better, sex.

In the wake of the women's movement, which challenged traditional notions of "femininity" (in part simply by pointing them out), researchers studying the psychology of cosmetic breast surgery shifted away from emphasizing the concept. As a review of the methods sections of articles examining the psychology of augmentation and reconstruction shows, however, the parameters of investigation remained generally the same: that is, they continued to focus on relationships, clothing, and the importance of appearance to self-esteem. Variables that previously had been linked by the claim to be measuring "femininity" now required another unifying concept. "Patient satisfaction" offered one such unifier. Rather than purporting to measure "objective" improvement in femininity, the new assessment strove to examine "subjective" improvement in satisfaction. Because this shift required no assumption of pathology and because it emphasized the power of the individual to assess the effects of surgery upon her own life, it fit well with the changing conceptualization of the implant patient as essentially normal.

A second subtext was the question of directionality. Did the need for implants precede the development of the technology, or was need one way in which the reach of that technology could be extended? There is no question that women's desire for breasts of the right size and shape existed. But the transformation of that desire to a medicalized need was something that followed the introduction of breast implant technology. "Need" allowed plastic surgeons to promote breast implants, to women and to their medical colleagues, as treatment for disease. In so doing, they established both the idea that such treatment was desirable and their own dominance as the gatekeepers of the technology. At first, need was important for providing a benefit with which to balance the risks of implants. Hypomastia and postmastectomy syndrome granted authority (an authority that was also highly gendered) to plastic surgeons; diagnosis bestowed legitimacy to the individual's suffering and to the plastic surgeon's treatment of that suffering. (Portraits of patients before and after: headless, handless.) When social changes—the women's movement, the widening commodification of the body, the self-actualization movement that emphasized "personal growth"—combined with a growing confidence in the technology, need was reconceptualized. The plastic surgeon–patient relationship became a partnership grounded in common aims. Cosmetic surgery became simply one route to helping the individual reach her "fullest potential."[150] (Portraits of women: outspoken advocates, gesturing.) Desire, in and of itself, became a sufficient need.

Although the controversy over the FDA's regulation of silicone gel breast implants turned around issues of risk, need was also an important issue, one that was embedded in discussions of benefit. The FDA and the proponents of continued availability agreed that the benefits of the device were psychological. While proponents argued that such benefits were self-evident, however, the FDA required data on such psychological constructs as "patient satisfaction, improved self-image, and improved outlook."[151] Implant manufacturers and plastic surgeons turned to the literature I have reviewed in this chapter, as well as to the testimony of psychiatrists, psychologists, and satisfied patients, to provide such data.

The imperative to produce proof of psychological benefit brought an interesting dilemma. On the one hand, an emphasis on the pathology of women seeking breast implants would serve to underscore the effectiveness of the devices, providing a benefit to balance the risk, as it had during the era of synthetic sponge implants. On the other hand, the particular context of the controversy militated for the portrayal of implant patients as psychologically healthy and able to make competent, informed choices about the devices. Implant proponents sought to strike a balance by hinting at larger problems but attributing to most patients only an extremely circumscribed psychological unease.

> Abnormally large or small breasts can lead to low self-esteem, little self-confidence, depression, and inability to seek an intimate relationship which leads to marriage and children, which are desired by many women. . . . Our literature review indicated that women choosing to undergo augmentation are healthy and well-adjusted. These women are not found on study to be neurotic, but to have feelings that they differ from other women only in a negative evaluation of their breasts. Augmented patients report improved self-esteem, improved body image, increased social expression, the ability to wear a wider range of clothing and a restored feeling of feminine identity.[152]

Implant proponents knew that the psychological benefits of reconstruction were accorded more legitimacy than were the benefits of augmentation. Thus many arguments about need sought to conflate the two: "The distinction between the psychological indications for [reconstruction and augmentation] . . . is not as clear as it might first appear. . . . the psychology of a woman's attitude towards a breast is deeply rooted and equally important in both circumstances. The psychological pain over a discrepancy between social and personal body image norms and one's actual body is real and significant pain."[153]

Since the motivations and benefits of the two procedures were identical, to draw a policy distinction between them would be "judgmental." As a female plastic surgeon who was also an implant recipient stated, "The opinion that women who have a medical reason to choose breast implants, whereas those of us who chose them for other reasons, psychological and emotional, are somehow unworthy, I find both judgmental and irrelevant."[154] The ASPRS made the same argument, claiming that "in light of the substantial benefits of breast implants, to view them as unnecessary or frivolous is to demean those women who feel physically handicapped whether by virtue of surgery, injury, congenital happenstance, or self-image."[155] Even Timmie Jean Lindsay, the first woman to receive silicone gel implants, spoke for their psychological value. Although she had been in the hospital for removal of a tattoo when Cronin and Gerow approached her to ask if she would be the test case for their new device ("join the team" is how Lindsay described it), Lindsay recalled that "the impact the implants had on me, and the restoration of my self-esteem gave me the courage to go out in the world to get a job, go to work, support my children and I was able to keep them with me and raise them through the years."[156]

The need for implants was captured in one statement: "Most women who choose implants just want to look normal."[157] "A sense of normalcy" was the goal both for reconstruction patients and for women seeking augmentation, who "do not identify themselves as merely having small breasts, but rather as having a deformity" and for whom surgery "plays a *restorative* and *compensatory* function" that allows them "to join the ranks of 'normal' women."[158] As we have seen, however, "normal" was a constantly shifting notion. The appearance of the normal breast changed with the fashions; the desire for breast surgery might be normal or pathological, depending on the broader context. Implicitly, throughout both the history of implant technology and the "psychology of the flat-chested woman," normal was not defined statistically or objectively, but individually: a matter of subjectivity that, in the end, could be understood only through empathy, not science.

5

REAL SUFFERING: *The Creation of Silicone Disease*

> All disease is socially created reality. What it means and the response it evokes have a history.
>
> —Ivan Illich, *Medical Nemesis*

The "diseases" for which implants were the cure were disorders of mind and body, psychological suffering triggered by a less than optimal appearance. Silicone disease—in a kind of inverted parallel—was a collection of wide-ranging signs and symptoms that sufferers and their supporters believed to have been caused by implants. Like hypomastia, silicone disease was "discovered" only when there was a reason to do so; that is, when it became necessary to establish the legitimacy of the damage and suffering caused by implants.

This chapter looks at the development of ideas about the harm caused by silicone implants. Harm was conceptualized first as a local "side effect," then—with the possibility of carcinogenesis and later of autoimmune disease—as systemic. The realization that silicone might cause systemic autoimmune problems arrived slowly and with it came a question: was silicone implicated in the etiology of several known, well-described disorders—or was it linked to another constellation of signs and symptoms, the manifestation of a new, silicone-specific disease? Acceptance of such a silicone-specific disease would not only explain the signs and symptoms of systemic damage but also legitimize the personal suffering of its victims. The stories of these victims portrayed their suffering as caused by something more than exposure to silicone; it was the result of perfidy and malfeasance. Thus victims of the disease required not only treatment for the body but legal and social remedies as well. The defenders of implants—mainly manufacturers, plastic surgeons, and satisfied implant recipients—who opposed these legal and social remedies, sought to prevent their implementation by disputing the reality of the disease.

Reports of the adverse effects of the early breast implant technologies—paraffin, synthetic sponges, and silicone injection—emphasized local damage like inflammation, hardening, infection, and scarring. Many, however, also noted the potential for, if not the fact of, systemic effects. In 1930, Karl-Heinrich Krohn described the systemic dispersion of injected paraffin, raising questions about toxicity and "connective tissue reaction." He reported several cases of women who exhibited problems of "feverish rheumatism" after injection. (Their symptoms abated, Krohn wrote, after surgery to remove all visible paraffin.)[1] John Grindlay and John Waugh, who tested sponges on animals in the early 1950s, alluded to the connection between the local and the systemic when they praised the "biocompatibility" of synthetic sponges and when they warned of the potential of the material to cause cancer.[2] (Proponents like W. John Pangman disputed this warning but clearly felt that they had to answer it.) In the 1960s and 1970s, the deaths caused by silicone injection were attributed to the liquid's problematic propensity to disperse throughout the body. Also, the increasing sophistication of the sciences of toxicology and immunology provoked the recognition that the chronic inflammatory response seen at the site of injection might be more than something local.

Debate over the natural and the artificial turned around the question of systemic reaction. Even proponents of synthetic implant materials recognized that it seemed to be "natural" for the body to mount a response. "It is always with reluctance that the plastic surgeon prefers an artificial implant to an autogenous transplant," noted one surgeon. "Irrespective of the qualities of the chosen material, he recognizes that only under the most favorable circumstances will the reticuloendothelial system [a system of immune response] remain vigilant and yet tolerant. When these benevolent circumstances cease to exist, a combat ensues which can lead to the rejection of the foreign body."[3]

The belief that silicone was "inert" came from the early observation that the material itself remained essentially unchanged when introduced into the body. Silicone was "water soluble; resistant to oxidation and chemical attack . . . generally non-volatile and noncrystalline."[4] Connective tissue grew around silicone but did not invade it. The claim of nonreactivity rested on the apparent insouciance with which the body accepted implants. Short-term tests of toxicity found that silicone caused no apparent harm in animals. Similarly, investigations by Dow Corning and the Eli Lilly Company into the biological activity of silicone—that is, tests to determine if it might hold promise as a therapeutic agent—showed no such effects.[5]

When, in 1982, the FDA announced its intention to place silicone gel breast implants in the regulatory classification that would require manufacturers to produce proof of their safety and effectiveness, the chief systemic concern cited by the agency was carcinogenesis. As discussed earlier, the possibility that implantation of synthetic materials might cause cancer in the "vulnerable" tissue of the breast had been a primary concern during the period when implants were being developed. Research with animals had focused on tumorigenesis. Plastic surgeons had conducted surveys and published case reports that looked at occurrences of postimplantation breast cancer. By 1982, because they were not seeing it in their practices, the possibility that silicone implants might be carcinogenic had ceased to be of any great concern to plastic surgeons. The FDA's notice, however, prompted them to refocus on the issue.

The FDA argued that "insufficient time has elapsed to permit the direct evaluation of the risk of cancer" from silicone implants, nor were there "sufficient epidemiologic data or experimental information in animals to make an inferential judgment."[6] In indicating that data on cancer risk would be required in order to ensure the continued availability of silicone implants, the FDA provided a road map for the manufacturers and plastic surgeons.

At the 1984 annual meeting of the American Association of Plastic Surgeons, a team including Dennis Deapen, an epidemiologist at the University of Southern California, and plastic surgeon Garry Brody presented research clearly designed to answer the FDA's demand for data. The Deapen study assessed the risk for cancer after augmentation surgery by linking the records of three thousand augmentation patients (culled from the patient records of thirty-five southern California–area plastic surgeons in private practice) with those of a statewide cancer registry. The rate of cancer among the augmentation group was then compared to the rate that would be expected among the general population. Results of the study showed that the incidence of breast cancer among women who had undergone augmentation was lower than one might expect to see in the population.[7]

Although the Deapen study had clear limitations—a comment appended to the published version noted the short follow-up times and the relative youth of the women studied[8]—plastic surgeons fastened upon it as a scientific demonstration of their clinical impressions. The results thus bolstered, in general, their trust in those experience-grounded observations and, specifically, their belief in the safety of implants. During the uproar following the 1988 release of the Dow Corning rat study, implant proponents used the Deapen study to argue that the findings in rats bore no relevance to the human population. Several years later, its results were reinforced when a Canadian study also found a lower than expected rate of breast cancer among a group of implant recipients.[9] (A follow-up to the Deapen research, published in 1992, continued to find a low risk for breast cancer among augmentation patients. This time, however, the authors reported what appeared to be a significantly elevated incidence of lung and vulvar cancers among women with breast implants.)[10]

The Deapen study did not seem to have the same power for the FDA, however. In its 1990 notice calling for applications from implant manufacturers, the agency used the Dow Corning rat study to justify its listing of carcinogenesis as one of the possible risks of silicone implants. In addition, the agency had added a new cancer-related risk: interference with early tumor detection.

The possibility that implants might impede breast cancer detection by obscuring breast tissue during mammography had received a fair amount of attention in the medical literature. A 1988 article published in *Plastic and Reconstructive Surgery* pointed to a history of dispute over the issue.[11] While one early publication had asserted that mammography was actually "facilitated in patients with severe hypoplasia . . . because of the postoperative protrusion of the mammary gland"[12] (an assertion repeated in the popular press,[13]) other reports found that silicone gel implants in fact interfered with mammography. (This interference occurred because the implant itself masked views of the breast tissue and compressed the breast tissue, making it denser and thus harder to read.) One author estimated that implants might block three-quarters of the breast tissue. The authors of the 1988 article estimated that, on average, more than a third of breast tissue could be obscured when implants were present. Given this estimate, they concluded that standard mammography was not a "reliable" screening method for women with implants and urged that new mammographic methods be developed specifically for this group.

Other research suggested that the mammographic problems caused by implants might be having a measurable effect on the diagnosis of breast cancer in women with implants. In one study women with implants were found to be in a later stage of cancer at the time of initial diagnosis than were women without implants. Nearly a quarter of cancers in the nonaugmented group were occult, but no occult cancers were found in women with implants.[14] Special techniques for positioning the breast tissue during mammography seemed to improve visualization, but because new views were meant to supplement, not replace, the traditional methods, women with augmented breasts would be exposed to higher total doses of radiation during mammography.

Although it was concern over autoimmune disease, not cancer, that was most prominent during the controversy over silicone gel breast implants, these investigations of cancer risk revealed certain themes that were to recur during the later debates. The focus on *breast* cancer—even after a study had shown an increased risk of cancer at other sites—showed the propensity of plastic surgeons to look only for "local side effects," ignoring evidence for the systemic dispersal of silicone. The collaborations that developed for the purposes of research—collaborations with radiologists and epidemiologists—mirrored the alliances formed with psychiatrists and pathologists and showed the willingness of plastic surgeons to borrow expertise when it suited their purposes. A single study that confirmed their clinical experience was enough for the plastic surgeons, but the FDA had a different way of weighing and assessing scientific evidence. This disjunction in approach to risk assessment was to be one of the points of tension during the controversy. Finally, the issue of mammography was important to the FDA advisory panel's eventual recommendation that the access to implants be more restricted for augmentation patients than for reconstruction patients.

The first reports linking breast augmentation procedures to immune disorders appeared in Japan in the early 1960s. A 1964 article described two women who had had paraffin injections for breast augmentation and later developed connective tissue disease with hypergammaglobulinemia (an indicator of a heightened state of immune response). The authors analogized this condition to an arthritis that could be experimentally induced in rats by injection of a mixture of mineral oil and heat-killed tubercle bacilli (called Freund's adjuvant), and named it "Human Adjuvant Disease" (HAD).[15] HAD was distinguished by several (awkwardly translated) characteristics: "1) Autoimmune diseaselike symptoms develop in women usually 2 years after plastic surgery with foreign-substance injection; 2) Silicone or related substances . . . possibly have adjuvant effects; 3) Foreign [body] granulomas can be observed histopathologically in an injected area and in their associated lymph node drainage; 4) The patients have some serologic abnormalities, such as autoantibodies; 5) The symptoms of some patients improve after the removal of foreign substances; 6) No infection or malignancy existed in the operated region."[16] Conceptually, the important features of HAD were the apparent exposure-response relationship—symptoms developed after initial exposure but resolved after the source of exposure was removed—and the linkage of local effects, like granuloma, and signs of abnormal systemic immune response, like the production of circulating autoantibodies.

More reports of HAD appeared in the Japanese medical literature in the 1970s. One article described seven women who showed signs and symptoms such as arthritis and arthralgia (painful joints) and laboratory values indicating a systemic immune response. Each woman had had injection mammaplasty with paraffin or liquid silicone.[17] Another article reported nine more cases, including four women diagnosed with scleroderma (an autoimmune disease characterized by swelling and thickening of the body's fibrous tissue), a disease that was rare in the general population. The authors of this article estimated that more than thirty cases of HAD linked to breast augmentation had appeared in the literature since the original description in the early 1960s.[18]

Although HAD provided a framework for understanding the adverse effects of both paraffin and silicone injections—effects that had been widely reported in the Western literature—the Japanese reports received little attention at the time. The inaccessibility of the journals in which they were published (most were not even abstracted in English) was one explanation. But so too was the general disclaimer, noted in chapter 3, that the injection method of augmentation was used only "someplace else." The cloak of disreputability served to mask the possible relevance of HAD to the legitimate uses of other silicone materials.

In 1982, several Australian authors reported the cases of three women who had come to them with, respectively, systemic lupus erythematosus, classic mixed connective tissue disease, and rheumatoid arthritis with Sjogren's syndrome (all well-known autoimmune disorders). Each woman had had breast augmentation with silicone gel implants about two to three years before the initial onset of her symptoms. Noting the clinical similarities between these cases and those reported in the Japanese literature, the authors suggested that they may have found several more examples of HAD.[19]

The specificity of the link to silicone gel breast implants and the diagnosis of well-known autoimmune diseases, as well as the imprimatur of Western practitioners writing in English in a Western medical journal, brought this article more attention than had been given to the earlier reports. Within five years several more reports of autoimmune disease in women who were silicone breast implant recipients appeared in the medical literature. (A rheumatologist later noted that it was at about this time that he and his colleagues began asking female patients presenting with muscle and joint pain if they had silicone implants.)[20] A review of these reports noted that patients seemed to fall into two distinct categories.[21] One group presented with classic, and easily recognizable, forms of connective tissue disease, in particular rheumatoid arthritis and scleroderma. Another group, however, demonstrated a range of symptoms and clinical indicators that were suggestive of autoimmune disorder, but which did not match any one disease entity.

Lawyers for Maria Stern, who in 1984 sued Dow Corning for product liability and fraud, used the Japanese reports and the Australian article, as well as the testimony of medical experts, to support their contention that Stern's autoimmune problems had been caused by leaky implants. Another source of evidence were the documents, uncovered in the course of pretrial discovery, that showed that Dow Corning had suppressed data about the possibility of adverse health effects caused by silicone, including one study in which dogs exposed to silicone became very ill and showed clinical signs indicative of an autoimmune disorder. Stern's victory—Dow Corning was ordered to pay compensatory and punitive damages totaling about $1.7 million—rested on the finding of possible suppression of this scientific evidence.[22] (However, as discussed in the introduction, the existence of this evidence was itself suppressed as part of the eventual out-of-court settlement between Stern and Dow Corning.)

The Stern case had a lasting significance for the conceptualization of silicone disease. Its verdict provided a form of legitimization for the connection between silicone and autoimmune disease, but the fact that this legitimization came in the courtroom—where a case is woven using many strands, and kinds, of evidence—meant that the issue of disease was inextricably tied up with the legal issue of malfeasance. The disease *was* the falsification and cover-up of scientific data; the disease *was* the failure of industry and plastic surgeons to inform women of the risks. For those who believed in the reality of silicone disease, this linkage would serve to strengthen their conviction. For those opposing that reality, however, the use of such evidence, and the site of legitimization, would lend credence to the argument that silicone disease was a wholly political diagnosis.

In 1986, a major review article on the subject of silicone implants and autoimmune disease appeared in *Plastic and Reconstructive*

Surgery, the journal of the ASPRS. The authors, who were plastic surgeons, after providing what they called an "immunologic primer"—basic information on the science of immunology—reviewed the clinical literature back to the early Japanese reports. Although "silicone is relatively inert chemically, and it is tempting to equate this inertness with biological inactivity," they warned, "it is becoming clear that silicone is not biologically or chemically inactive."[23]

There were four major hypotheses about the etiologic mechanism of silicone-linked immune disorders.[24] The adjuvant hypothesis posited that silicone was a sensitizer that boosted immune response to other antigens. The hapten hypothesis suggested that silicone bound with naturally occurring substances in the body, changing their conformation so as to render them "other," thus triggering an immune response. A third hypothesis suggested that silicone itself was an antigen that, in sufficient quantity or the right chemical form—particularly the short chain of liquid silicone—could provoke an immune response. The fourth was that silicone in the body could be converted into silica, a compound of silicon and oxygen that was known to be highly antigenic. The idea that individual susceptibility played a role in the development of autoimmune disease cut across all four hypotheses.

Backers of each hypothesis could claim some support in what was known about silicone, but there was no conclusive evidence. Determining which theory captured what—if anything—was going on would require more study. As the authors of the *Plastic and Reconstructive Surgery* review article argued, such research would have to be interdisciplinary, encompassing epidemiology, animal experimentation, and clinical immunology. "We must remember that these reports will not go away if we ignore or deny them," urged the authors. "It is incumbent upon us to approach the problem scientifically."[25]

In their statements to the FDA and to the public, however, both the plastic surgeons—as represented by the ASPRS—and the manufacturers spent the next five years arguing that the problem of autoimmune disorders linked to silicone exposure was a non-issue, unsupported by any scientific data. This public stance reflected both the surgeons' clinical experience—they simply were not seeing such problems in their practices—and the conflict over evidence and the interpretation of evidence that was embedded in the broader controversy over silicone breast implants.

In the 1990 notice that announced the FDA's intention to require manufacturers to provide proof of the safety and effectiveness of implants, the agency, for the first time, indicated its concern about the possibility that implants caused immunological sensitization leading to autoimmune disease. To support this concern, the notice cited the case reports of the 1980s that had linked implants with autoimmune disease, a laboratory study on the biocompatibility of silicone, and an early report by plastic surgery investigators showing that the capsules that formed around breast implants contained particles of silicone that had been shed from the implant envelope. The FDA warned that the manufacturers would not receive agency approval unless they addressed these concerns by providing "detailed discussion with results of preclinical and clinical studies of the risks identified."[26]

By early 1991, when the FDA sponsored the conference "Silicone in Medical Devices," there was enough research going on to support more than twenty presentations (and numerous poster displays) on the chemistry of silicone, the biocompatibility of the material, the toxicology of free silicone in the body, the clinical manifestations linked to silicone exposure, and the particular complications of silicone breast implants. As the FDA expressed in its summary of the conference findings, however, scientific opinion remained mixed: "Some experts concluded that these devices have little, if any, serious adverse effects. Others claim that breast implants may pose a number of health risks including the potential to cause arthritis-like problems if small amounts of silicone leak from implanted devices and migrate to other parts of the body."[27]

This expert uncertainty, however, had long been rejected by the women who believed themselves to be ill because of implants. During the period between 1988 and 1992, the FDA received hundreds of letters from such women. These letters described local complications—breast pain (reported by 40 percent of the women who wrote to the agency), rupture (31 percent), and capsular contracture (29 percent)—and systemic symptoms, including joint pain (39 percent), fatigue (35 percent), and muscle pain (13 percent). For those women whose letters mentioned that they had received diagnoses, the most common were arthritis (19 percent), (unspecified) autoimmune disease (7 percent), Raynaud's disease (5 percent), and (also unspecified) connective tissue disorders (4 percent).[28]

Despite the disparity of symptoms and diagnoses, these women spoke with a common voice. An analysis of 112 letters received by the FDA during the first six months of 1992 found that they reflected four themes: "Most notable was the consistency with which women described the lack of information given to them, prior to surgery, about potential problems that could arise from implants. The women expressed anger that they had chosen breast implants without being fully informed. . . . The second commonly discussed theme was persistence of pain or other symptoms that were often reported as not taken seriously by physicians or not considered to be related to the implants. . . . The third theme [was] loss of ability to work or carry on normal activities. . . . The fourth theme [was] concerns about future problems and the inability to get information."[29]

More detail could be found in newsletters of silicone support organizations. An edition of the Command Trust Network (CTN) newsletter published during the winter of 1989–90, for example, summarized the diagnoses that had been reported in the medical literature: connective tissue disorders and autoimmune problems like arthritis, Raynaud's disease, scleroderma, lupus, and Sjogren's syndrome. The symptoms associated with these "painful, crippling, life-shortening diseases" included joint pain, stiffness, swelling, chronic fatigue, skin thickening and tightening, fibrotic hardening and dysfunction of organs and tissues, lung and breathing problems, rashes, sensitivity to light and cold, dry and burning eyes, mouth, and nose, and vision problems.

But CTN also described problems that had not yet appeared in the medical literature, such as weakness and pain in the shoulder, arm, and hand; sensations of skin burning; drenching sweats and chills; neurologic symptoms like memory loss and chronic headaches; gastroenterologic symptoms like nausea, vomiting, and irritable bowel syndrome; gynecological problems including infertility, miscarriage, and stillbirth; and cardiovascular problems including heart palpitations and stroke. In addition, the editors noted the severe emotional toll of psychological problems caused by the lack of effective medical help for these symptoms. (In a later newsletter, published after the FDA had announced its implant policy and in the midst of the legal wrangling over a financial settlement by the implant manufacturers, the list of symptoms had grown, approximately doubling in number to cover two singled-spaced pages.) The implication of this list of symptoms—and the explicit conclusion reached by many silicone victims—was they were experiencing something other than run-of-the-mill classic autoimmune disorders: their bodies were manifesting a new silicone-specific entity.

In *The Untold Truth,* a book about implants written by and for women who believed that they were victims of silicone, the authors attempted to convey a sense of the "real suffering" caused by this new disease.

> Some women are completely disabled; stricken down in the prime of their lives, and remain prisoners within their own bodies. Some are blinded, some are paralyzed, others are losing family, friends, and support systems because the strain of chronic illnesses is just devastating. Still others are suffering seizures, strokes, and progressive diseases. Money for children's college education may no longer be there. Life savings have been completely depleted for many. Some have had the traumatic experience of losing their homes because of astronomical medical bills. These symptoms have altered lifestyles of entire families, taking emotional tolls on all members. This type of atrocity should never have been allowed to happen.[30]

One autobiographical account provides an illustration: "In 1976, when I had my first set of breast implants put in, I never ever thought I would be headed down a path where there may be no return! I only wish that someone else could have forewarned me on what was about to happen in my life. I had just turned 18 years old. The surgery was a birthday present from my parents. A birthday present I would never forget for the rest of my life."

Over the next four years the narrator had a series of problems with her implants, followed by several replacement surgeries. One pair of implants ruptured. Her suffering increased.

> Some of my symptoms included, (not in any particular order) firm breasts changing into rock hard breasts; my breasts created

> great pain for me and constant pain in the rib area; hair loss; lost my eyelashes completely (several times); fevers (that could last up to 12 days); sore throats; swallowing problems; easy fatigue; bruised really easily; rashes like hives; very thirsty; dry mouth and loss of taste; positive ANA [a measure of autoimmune sensitization]; progressive weakness (this was getting bad) including my limbs. They would tingle (pins and needles feeling). My arms/legs felt like rubber or numb and would last up to 3 1/2 hours (could not walk, or walked with a limp.) Memory loss or confused; shoulder pain; joint and bone pain—like hot rushes in my arms, legs and spine; swelling so bad I had to have my rings cut off. My eyes hurt, white part of my eyes were bright pink color and I felt like my eyes would drop out on the floor; dry eyes—I used artificial tears. Miscarriages/fertility problems; horrible periods that would last up to eight days with bad menstrual cramps and clots the size of my fist. Colds "all the time" where I would cough up yellow/green/brown sometimes red blood. I felt I was drowning and I would carry a cough for a long time; horrible chest pain associated with a cold or infection. Horrible headaches (not a lot) but when I got a headache it was bad; sickness/rashes from the sun/heat; breathing problems (like I couldn't catch my breath or my breathing was off); highs and lows (depression or great sadness). I felt as though I could jump out of my skin; irritable bowel. . . . Sometimes on my breast I would have some discoloration which always would resolve gradually but did not evolve through the greenish and yellow discolorations that was typical of bruises—this also happened on the inside of my arms; breast drainage—clear to yellow urine color—greenish plugs in the nipples; yellow breast milk; purple/white pasty hands when I went out in the cold weather; felt like I was getting the flu and other symptoms. To say the least I had my good days and bad days! All of this was like the domino theory. If you had one symptom the others were sure to follow the longer the breast implants remained inside your body. Also I lost faith in the doctors/medical field. They did not have any compassion or understanding for what I was dealing with or going through.

Eventually, she realized: "I was having a lot of the same symptoms as other women I had read about in the magazines and seen on TV. Coincidence—NO WAY! I knew I was getting a lot worse. I had to take action. . . . I had to stop the denial. These devices were my problems!" She had surgery to remove the implants and what loose silicone could be found, then experienced a remission of most of her symptoms: "It's my miracle! . . . I'm my old self again. Just like I was before the implants were put in. It seems as though the past 17 years were only a childish nightmare."[31] (In her breathlessness I hear an unconscious parody of the prose in women's magazines. Perhaps this is what the silent voice of Baker's "movie girl," whose augmentation surgery lives on in film, might have sounded like.)

While the scientific classification of disease necessitates a process of specification, a narrowing-down to elucidate a relationship of cause and effect, victims' conceptualizations of silicone disease tended to be expansive. Symptoms colonized every system of the body. They were unpredictable except for the predictability of horror. The harm caused by silicone, like a stone tossed into still water, rippled out: the body, the mind, the disrupted family, loss of income, loss of faith. The victims' attribution of causality was similarly expansive. Silicone disease was caused by silicone but also by deception, a lack of information sweetened by assurances of safety.

The erstwhile movie girl's experience of self-realization through news accounts of other women's suffering was one that became increasingly common as the controversy attracted more public attention and more stories about the possibility of silicone-linked harm appeared in the media.[32] Women's magazines, for example, which for years had promoted implants as an uncomplicated and safe tool of self-fulfillment, shifted their coverage to emphasize the possible dangers of implants, presenting accounts of women suffering from silicone-linked health problems and urging women to be alert to any signs of harm.

In *The Silicone Breast Implant Story*, an account of the breast implant controversy built around analyses of controversy-related narratives, authors Marsha L. Vanderford and David H. Smith examined how the news media reported silicone disease. Media stories focused on graphic representations of ruptures and leaks, implying that these occurrences were caused by negligence on the part of the manufacturers and that they were causally related to the development of autoimmune disease. "The narrative structure that emerged . . . takes the form of a horror story in which the heroine acts to solve a problem . . . encountering fearful challenges and difficulties along the way," Vanderford and Smith wrote.[33] "In these [horror] stories, the heroine begins confidently, trusting her plastic surgeon and his reassurances. Following the implant surgery, the woman experiences pain, mutilation, and unexplained illnesses. Trying to find relief, the heroine finds that people and events are not what they appear to be. Once-trusted physicians betray heroines by withholding support and critical information. The dream of a new figure fades in the realization of medical problems. Ultimately, fear and suspicion overcome faith. At the end of the horror story, the central character emerges sadder but wiser, often without her implants."[34] Apparently, the silicone victims' representation of reality—the conflation of physical symptoms and betrayal—made a very good story.

Perhaps the most widely viewed example of the "horror story" was the CBS broadcast of Connie Chung's *Face to Face* that aired in December of 1990. Chung interviewed several women with health problems they believed were related to their silicone implants, showed pictures of leaky implants, and spoke to several medical experts, one of whom was willing to implicate silicone as the direct cause of the women's immune disorders. The Weiss hearing, which took place just a week after the Chung broadcast and also received heavy coverage from the press, used the same narrative.

As media attention raised awareness of the possible connection between silicone and autoimmune disease, more women with implants—even those who had been asymptomatic—began to fear the devices. A letter to the editor of *Plastic and Reconstructive Surgery* characterized the media coverage as a barrage of "negative, usually one-sided publicity," estimating that between 1990 and 1991 more than two thousand articles or television reports on silicone breast implants had appeared and that three-fourths of them were negative.[35] Plastic surgeons blamed the media when they found themselves "inundated with inquiries from concerned patients."[36] Prompted by what they saw as an unnecessary degree of "hysteria" among these women, and by the belief that there existed a "silent majority" of breast implant recipients who were not experiencing problems, plastic surgeons launched a dual effort to study the effect of the coverage on their patients and to harness the power of the media for their own ends. (See chapter 6 for a discussion of the plastic surgeons' media campaign.)

In their research the plastic surgeons focused on the questions of which media sources were most influential, how much credibility women gave these sources, what information they gained from media accounts, and how what they learned affected their attitudes toward implants. Television turned out to be women's chief source of information about implants.[37] While women initially expressed some skepticism toward the accuracy of television coverage—many used words like "biased" or "sensational"—as the controversy progressed women seemed to lend more credence to what they saw. After January 1992, when David Kessler announced the moratorium on implant sales, women's confidence in the media increased, their concern about implants went up, and, for women who already had the devices, their satisfaction decreased.[38] One study, conducted in 1991, found that the two adverse effects women were most aware of were interference with mammography and capsular contracture,[39] but after the moratorium women (in another study) were far more likely to recall hearing reports of links between silicone implants and problems of rupture and leakage, the dispersal of silicone throughout the body, and autoimmune disease.[40]

When women trusted the media more, they trusted plastic surgeons less. An ASPRS survey conducted after the announcement of the moratorium found that only 64 percent of women in the general population thought that plastic surgeons were trustworthy sources of information about implants.[41] As media "horror stories" achieved credibility, so too did silicone disease.

Just as silicone victims tended to conflate the suffering caused by the physiological effects of silicone with the suffering caused by the perceived betrayals perpetrated by plastic surgeons and manufacturers, so the proponents of implants saw as inextricable the tie between silicone disease and the "politicization of implants." The problem of politicization was understood as one of the dominance of consumer claims over scientific claims. Every time the media publicized these consumer claims, silicone disease was recognized, recognition that both sides of the issue saw as a step toward legitimization.

One of the issues at stake in the controversy over the FDA's regulation of silicone breast implants was this question of the legitimacy of silicone disease. For women claiming to have been harmed by implants, and for their supporters, the struggle was to establish the reality of silicone disease as a clinical entity caused by exposure to silicone from defective implants. This reality turned on several issues: acknowledgment of individual and collective suffering; the denial of true informed consent; and the perfidy of plastic surgeons and manufacturers in promoting their own interests above those of women. Inducing the FDA to ban or severely restrict the future availability of silicone gel breast implants would serve this view of reality because, by implication, such a policy move would acknowledge the dangers of the devices. In contrast, proponents of implants sought to maintain the widespread availability of the devices by denying the reality of silicone disease. For plastic surgeons, manufacturers, and women who were satisfied with their implants and had been spurred to advocacy by the threat to the devices, silicone disease became a symbol not of harm or perfidy but of restriction on personal (and professional) freedom, of greed, and of "pathologic science." Implant proponents sought to refocus the debate around breast cancer, a disease whose reality had been firmly established medically, scientifically, and, most recently, politically.[42]

A 1992 analysis of the media coverage of the controversy decried a tendency to frame it as an issue of conflict between opposing groups of women—the silicone victims and the advocates for implants who were also satisfied recipients—rather than as one of serious corporate (and perhaps government) misconduct.[43] A closer look at the dispute between the victims and the advocates, as represented by their testimony during the Weiss hearing in 1990 and the November 1991 FDA advisory panel meeting, is useful here, however, because it provides a window on their alternate realities in the construction of silicone disease.

Both groups claimed that science supported their positions. Implant advocates arguing that the devices were safe pointed to the paucity of any "good science"—epidemiology, laboratory experiments—that proved harm. They dismissed case reports as anecdotal. The invidious juxtaposition of "good science" and "anecdote" enraged silicone victims, who interpreted it as an attack on their own credibility. Their response was to present their own stories of suffering as refutation. How can you call what is happening to me anecdotal? they asked; I am the proof, and my pain is the evidence. Silicone victims accepted case reports as science, pointing to a tradition in which such reports were often the first sign of serious environmental risks.[44] They lauded the work of "our doctors"—physicians known to be willing to treat silicone victims and to be sympathetic to their concerns—and argued that their proposed hypotheses of the mechanisms of harm were groundbreaking science which eventually would reveal the truth.

The two groups also presented divergent conceptualizations of risk. The silicone victims argued for zero tolerance. CTN, for example, called for a ban on implants "until we are absolutely certain that the safety of these implants can be proved beyond a reasonable doubt."[45] Implant advocates saw risk as existing in a balance with benefit. As they extolled the value of implants for reconstruction and for raising self-esteem, they were arguing for allowing a greater potential risk. Life itself is high risk, they asserted; government regulation cannot insulate individuals from all potential harm. Do not ban implants, an advocate told the FDA advisory panel, "unless you can scientifically prove that implants are the direct cause of life threatening illness."[46]

As the advocates' argument about risk hints, the two groups took different positions on individual responsibility. While the silicone victims—even those who claimed to have done extensive "research" before choosing to have implants—blamed the perfidy of doctors for failing to warn them of potential problems, advocates portrayed themselves as completely informed of all possible risks. Even those women who had experienced complications claimed they accepted responsibility for the choice. Linked to their assertions of individual responsibility were expressions of confidence in their doctors: their doctors kept up with the latest information; their doctors told them the truth. (By implication, then, they suggested that many of the victims' problems had been caused by bad doctors, individual surgeons who might be sued without endangering implant access for everyone else.) Breast cancer patients know there are "trade-offs and uncertainties" with any medical procedure, said one implant advocate, implying that silicone victims were augmentation patients who either had ignored the uncertainty when they chose implants or were doing so now in order to strengthen their legal cases.

Both groups marshaled feminist arguments to support their causes. Silicone victims pointed to a history of inadequate research and technological abuse in women's health, analogizing the breast implant story to that of DES or the Dalkon Shield. They emphasized the lack of information in the informed consent process, arguing, as one representative of a women's health organization put it, that "informed consent is only as good as the information available to those making the decisions."[47] Implant advocates, consistent with their focus on personal responsibility, emphasized the consent. Women must be viewed as competent adults in matters of medical decision making, they asserted; to do otherwise would be a form of paternalism. In addition, the advocates raised objections to the FDA hearings which they couched in feminist terms: Why was the panel not considering devices used by men? they asked. Testicular implants were nearly identical in design to breast implants, they argued, yet no one was threatening to take them off the market. Several women implied that the hearings were an exercise in voyeurism, an excuse to talk about women's breasts in public. One woman noted her resentment at having to justify her choice and said that she considered it a form of sexual harassment.[48]

Perhaps the most striking difference between the two groups was their perceptions of each other. Implant advocates perceived the victims as a vocal minority, the "4% who are trying to penalize the rest of us." They saw the victims' illnesses as a matter of coincidence or blamed their difficulties on incompetent surgeons. Some minimized the physical problems described by the victims, arguing that they too suffered "arthritic symptoms," but noting that such problems were inevitable with aging and could be controlled with medication. Individual advocates accused women who were now complaining of harm of exaggerating their suffering in order to improve their chances of winning large legal settlements. Several advocates spoke with resentment of the claim that support groups like CTN made to represent them. *They* don't speak for *me*, these women countered. In all, the women advocating for the continued availability of implants tended to set themselves apart from the silicone victims.

In contrast, the silicone victims looked upon the advocates as sisters who, so far, had been luckier than themselves. What separated the two groups was the immediacy of their suffering. We are "the tip of the iceberg," they claimed. Silicone was a ticking time bomb that would explode inside women who were still healthy. The claim to represent all women with implants was based on their belief in the inevitability of harm. (Questioning by a member of the FDA panel revealed another important difference between the two groups that lent credence to this argument: among those testifying against implants, the average length of time since implantation was nine years, while those women speaking in support had had them for an average of only about four and a half years.)[49]

The mandate of the November 1991 FDA advisory panel was to rule on the sufficiency of information in the manufacturers' applications. As discussed in chapter 1, the panel used this charge to sidestep any explicit consideration of implant safety. Thus they managed to avoid making any statement about the reality of silicone disease. Before the year was out, however, the legal system showed itself to be willing, once again, to recognize, and compensate for, the health problems linked to silicone.

In *Hopkins v. Dow Corning* a jury accepted the plaintiff's claims that she had been harmed by defective implants and that the manufacturer had committed fraud because it had known of the defect and of the likelihood of harm, but had covered up its knowledge. Among the evidence presented in the trial were the Dow Corning documents, which revealed that the company had been aware of the possibility that silicone caused autoimmune disease–like problems in dogs. The jury's decision legitimized the silicone victims' construction of the problem of implants as one of suffering and perfidy. It also bestowed legitimacy on the hypotheses of the doctors and scientists who testified about mechanisms by which silicone might cause harm. By legitimizing the implant opponents' construction of silicone disease—especially in the glare of publicity now attendant on silicone breast implants—the court ensured that the problem, and the relevant evidence, would have to be considered by the FDA.

When FDA commissioner David Kessler ignored his own advisory panel's recommendation—the November panel had recommended that implants remain widely available while more information about safety was collected—and instead called for a moratorium on the sale and distribution of the devices, he cited "new information" that had "increased [the agency's] doubts" about the safety of implants. The information he described focused on the issue of quality control problems and the possibility that silicone might be linked to systemic illness. Kessler made it clear that the focus of the upcoming advisory panel meeting would be on the question of autoimmune disease.[50]

Implant defenders reacted to Kessler's announcement by objecting to the source of the "new information." Their rebuttals featured claims that the new evidence was political, not scientific. In its report on the Dow Corning documents, for example, the ASPRS argued that there was a "very great difficulty inherent in arriving at any scientific conclusions based upon review of information selected by attorneys to prove an allegation. *Science* is the gathering of ALL evidence before attempting a conclusion."[51] In drawing this sharp epistemological dichotomy, the plastic surgeons implied that silicone disease could not be a medical-scientific entity because it was a legal-political creation.

When the FDA's advisory panel convened in February of 1992—its membership revised so as to reflect both Norman Anderson's role in bringing the Dow Corning documents to public attention and the new focus on autoimmune disease—agency staffers laid out the "new information" and the questions it had raised. They focused first on the issue of quality control as it related to implant rupture and gel bleed and then on how exposure to silicone might cause immunological sensitization. Evidence for the possible adverse effects of silicone came from clinical observation and laboratory studies. FDA staffers reviewed the Japanese history and the formulation of HAD. They cited the case reports that had linked implants and immune disorders. They described studies that had shown the presence of low-molecular-weight chemical entities in silicone gel, noting the known connection between these low-molecular-weight products and immune activity. One staffer presented the results of an FDA survey of rheumatologists, whose consensus was "1) there is a possible association between silicone gel–filled breast implantation and rheumatic or autoimmune disorders; 2) the onset of the disease may be many years after the implantation, although acute cases have occurred; and 3) there is a need for a large epidemiologic survey to scientifically determine the incidence of the problem."[52]

The FDA also took up another problem. The November hearing had shown that the manufacturers had no precise information about the total number of women who had received implants. The oft-repeated estimate of two million women was based on manufacturers' sales figures and ASPRS tracking of the number of procedures performed by its members, but because it was common for one woman to have repeat surgeries in which she received several implants and for surgeons who were not ASPRS members (or even plastic surgeons) to implant the devices, the number was not reliable. In the February hearing, an FDA staffer presented the results of a recent population-based survey suggesting that the actual number was much smaller than the common estimate—not more than one million.[53] The importance of the revised estimate lay in its utility as a possible denominator. Any epidemiologic assessment—incidence, prevalence, relative risk—requires a denominator. That is, to assess the danger of an exposure, one has to know the total number of people exposed, as well as the number who became ill. The downward revision of the estimate increased the significance—both statistical and conceptual—of the reported cases of autoimmune disease.

In addition to the presentations by its own staffers, the FDA provided testimony from several clinicians and laboratory scientists. The clinicians included rheumatologists and a neurologist, all of whom had published articles about silicone-linked disease. These witnesses described distinct constellations of signs and symptoms seen in patients with implants, abnormal test results suggesting immune activation, and a pattern of improvement after implants had been removed. The laboratory scientists described studies showing that women with implants had the same patterns of autoantibodies as did patients diagnosed with certain immune disorders. Other studies had detected the presence of a silicone-specific antibody in several symptomatic implant recipients. Together, the clinicians and the laboratory scientists suggested that what they were seeing was not a well-described clinical entity like scleroderma or rheumatoid arthritis—they agreed that there was not enough epidemiologic evidence to support the contention that silicone caused these disorders—but rather, as the silicone victims had been arguing, a new silicone-specific disease, one with a latency period of about nine years after implantation.

The FDA's evidence was disputed vigorously by the manufacturers and the plastic surgeons. All manufacturers claimed that their extensive animal testing of silicone had never produced any results suggesting that it could cause immune problems. They questioned the clinical evidence, as well. Evidence from cases of silicone injection—as in the Japanese cases—was irrelevant. Silicone was ubiquitous in the modern environment. Antibodies to silicone had been found in people without implants. There was no epidemic of autoimmune disorders among insulin-dependent diabetics, who received a dose of liquid silicone every time they used a syringe. The occurrence of autoimmune disease in women with implants was a coincidence: the number of cases reported did not exceed the number one might expect to see in a population of the same sex and age. The mechanisms proposed to explain the process of immune activation had not been proven. Only epidemiologic studies could provide the data needed to resolve the question of causation, but such studies would be difficult to perform.

In a review of the literature on silicone-linked autoimmune disorders, the ASPRS argued that the FDA's concern "is neither supported by the numerous studies it cites, which fail to show that silicone causes morbidity or mortality, nor the generally accepted understanding of the inert character of medical grade silicone."[54] The review "concluded that at present there are insufficient data to suggest more than a speculative association between silicone gel–filled implants and the development of autoimmune connective tissue disease. . . . only carefully designed epidemiologic studies can determine the presence of and quantify any risk attributable to silicone gel-filled implants. . . . the concept of excess risk is speculative at most and cannot reasonably be considered in determining whether these implants should be removed from general use."[55]

Since the formulation of Koch's postulates in the late nineteenth century, scientists have tried to develop criteria by which to establish disease causation. Such criteria require that exposure precede disease onset, that there be a correlation between the amount of exposure and the severity of disease, that the relationship between the exposure and the disease be both specific and consistent, and that the theory of causation be congruent with existing scientific knowledge.[56] Implant proponents' attempts to demolish their adversaries' arguments implicitly drew upon these criteria. That is, they questioned the timing of symptom onset in some of the more highly publicized cases, pointed out the paucity of evidence for any kind of dose-response relationship, disputed specificity and consistency by noting the ubiquity of silicone and the wide range of symptoms and diagnoses claimed by women, and rejected all of the mechanisms of immunological sensitization that had been proposed in the literature.

For the defenders of implants, the "evidence" used to support the connection between implants and autoimmune disorders was a manifestation of "pathologic science," or "errors in science created by a loss of objectivity, the science of the unreal and bias."[57] Such "junk science" came not from the laboratory but from courtroom testimony by paid experts, who distorted the truth in order to win large jury awards.[58] The FDA, by allowing this evidence to be presented, had legitimized it, thus falling victim to the pathology and betraying its own lack of scientific competence. "To show that silicone was a cause of autoimmune disease," one plastic surgeon patiently explained, "one would have to produce evidence that the incidence of these diseases is statistically higher in women with breast implants than in the normal [sic] population. One would also have to eliminate other variables that could be causative. Any study that failed to meet these criteria is not science, but conjecture. And speculation must be further discounted when those who propose it stand to profit."[59]

Like the proponents of implants, the silicone victims and other opponents of implants recognized that the FDA's focus on autoimmune disease had legitimized the problem. These groups, however, welcomed such legitimization as it validated their own long-held and deeply felt constructions of reality.

The FDA's February 1992 advisory panel issued no definitive statement about the link between implants and autoimmune disease, although it did recommend that women who were experiencing symptoms such as muscle and joint pain see their doctors and that more research, specifically epidemiologic studies, be conducted. In its decision—to restrict implant availability to clinical trials with differing access for augmentation and reconstruction patients—however, it is clear that the panel, while perhaps not convinced of the reality of silicone disease, had at least become sensitized to the possibility of its existence. The policy recommended by the panel, and later accepted and promulgated by the FDA, explicitly attempted to balance the effort to gather more information with the need to "limit those women who will be placed at risk," thus suggesting that, indeed, there was a risk.

Not surprisingly, the policy was widely interpreted as a validation of the reality of silicone disease. In the months and years that followed the FDA's decision, that validation became the impetus both to scientific study and to extensive legal wrangling. As the focus on silicone intensified, the disease was, paradoxically, both codified and discredited. (The post-1992 legal and scientific events are summarized in the epilogue.)

The creation of silicone disease was a process of recognition and legitimization. First, the conceptualization of harm was broadened beyond local side effects to encompass systemic damage. Then both local and systemic signs and symptoms were interpreted as indicative of disease. At first, these diseases were thought to be classic autoimmune and connective tissue diseases. Later, however, as the lists of reported signs and symptoms expanded, and as the paucity of epidemiologic evidence was emphasized, the signs and symptoms were reinterpreted as a new syndrome—silicone disease, a physical manifestation of both the toxicity of silicone and the social circumstances of its use.

The interests of the stakeholders in the implant issue were embedded in the struggle over silicone disease. For women who identified themselves as victims of silicone, the disease classification was a way to make their suffering meaningful, and also to increase their chances for financial compensation. For their supporters—scientists and physicians who took their complaints seriously—silicone disease was an exemplar of a new paradigm, an expansion of accepted models of disease causation to encompass the risks of new environmental exposures. As potential expert witnesses, these supporters also had clear financial interests in silicone disease. The FDA emphasized the possibility of autoimmune disease in the context of its own working epistemology and in the midst of a threat to its authority (see discussion in chapter 7).

For groups on the other side of the implant controversy, opposition to the legitimization of silicone disease was necessary to protect their own interests. Manufacturers disputed the disease because to do otherwise would be seen as an admission of malfeasance, an admission that would open them up to huge liability exposure. Plastic surgeons, too, reacted in part because of liability—that is, financial—concerns. For the plastic surgeons, resistance to silicone disease also represented an assertion of their own professional autonomy and a defense of the discipline's working epistemology. Finally, for the women who stepped into the role of silicone advocates, disbelief in silicone disease reinforced both their right to make their own choices—choice of surgery, choice of surgeon—and their conviction that these choices had been correct.

Once, meaning to make a note about "diseases as inventions," I instead wrote "diseases as interventions." Diseases are inventions because they are—both literally and figuratively—socially constructed. That is, they are recognized and legitimized through the acts of human beings. Diseases are interventions because as constructions they reflect the dominant interests of the times in which they are prominent.

Emily Martin, writing of the cultural significance of immunological thinking at the end of this century, noted "the emergence of the immune system as a field in terms of which all manner of questions and definitions about health are given meaning and measured."[60] It is no coincidence that silicone disease was first conceptualized at about the same time as AIDS was being identified and described. Rather, both developments are linked to the increasing sophistication of the science of immunology and the construction of the immune system as central to the body and its state of being. Politically, silicone disease depended on AIDS. The horror of wasted flesh implied in a whispered explanation of "acquired immune deficiency syndrome" was easily conflated with the explanation of "autoimmune dysfunction" and the tears of silicone victims. More practically, the success of AIDS activists in changing the FDA's procedures for approval of new drugs opened the agency to other challenges.

But silicone disease is a particular kind of illness, a manifestation both of chemical toxicity and of perfidy. As such, it has less in common with AIDS and more with several other illnesses that have arisen in the late twentieth century. For example, sufferers of multiple chemical sensitivity (MCS) and victims of toxic waste dumps tell similar stories of malfeasance, corporate and government cover-up, and skepticism and dismissal by medical professionals.[61] Their constellations of signs and symptoms are as wide-ranging as those of the silicone victims, their exposures as controversial. In bringing meaning to their suffering, both groups "join science with biography."[62] For the silicone victims, that meant their illness narratives became moral and political accounts, indictments of the powerful institutions they blamed for their problems. The courtroom and the media, rather than the examining room or the laboratory, are the sites of legitimization for these diseases. The challenge they pose to experts, and thus to the political status quo, becomes their defining feature.

6

THE SPECIALTY NEAREST SCULPTURE IN THE LIVING

Plastic surgeons have been central characters in this story so far and have appeared in many guises: tinkers, skilled hands, father figures, self-promoters, television stars who meet bitter ends. How to reconcile these many roles? This chapter explores the historical and epistemological background of the profession to understand why plastic surgeons acted and reacted in the ways that they did during the breast implant controversy. [1] First, it presents a short history of plastic surgery as a professional discipline.[2] The focus is on several themes: the development and dissemination of plastic surgical knowledge; the centrality of the surgeon-patient relationship to the theory and practice of the specialty; and the dual nature of plastic surgery (as both art and science). The second part of the chapter examines some of the specific productions (to return to the theatrical metaphor of chapter 1) mounted by the plastic surgeons during the controversy, in particular, educational materials, testimony, and media campaigns. As we will see, these productions were consistent with, and made understandable by, the history of the profession.[3]

The term "plastick" surgery was coined in 1818 by a German surgeon, Carl Ferdinand von Graefe. The word derived from the Greek *plastikos*, meaning "to form" or "to mold," and also, in German translation, "to sculpt." Plastic surgeons used the technical skills of modern surgery to sculpt human flesh; the word "plastic" connoted not artificiality but a melding of science and art.

The practice is ancient. Egyptian records, dating to 3500 B.C.E., refer to the surgical repair of mutilated facial features. Galen described tissue grafting around the turn of the first century C.E. The history of rhinoplasty is particularly well documented. Repair of missing or deformed noses dates back to the seventh century B.C.E., in India, where amputation of the nose was punishment for a variety of transgressions. Rather than bear such a stigma, those who had been so punished sought to conceal their disgrace. Indian surgeons developed a procedure that cut the skin of the forehead or cheek, then twisted it on a pedicle and swung around to shape a new nose. Another technique for nasal reconstruction was devised by the Branca family in fifteenth-century Italy. Their method cut a flap in the upper arm and tied the stump of the nose to it for fifteen or twenty days, until the tissues had bonded. (Illustrations from the work of Tagliacozzi, a sixteenth-century Italian known as the "father of plastic surgery," show the leather-and-metal contraptions used to immobilize the patient and hold the nose to the limb.) The flesh was then cut away from the arm and used to fashion a nose. Early surgeons also built nasal prostheses out of a variety of materials, including leather, silver, and gold. Astronomer Tycho Brahe wore a metal nose after he lost his own in a youthful duel.[4]

In the nineteenth century, with the advent of modern surgical techniques, plastic surgery flourished. Anesthesia, introduced in the middle of the century, allowed longer and more intricate surgeries. Antisepsis and asepsis, popularized in the 1870s, improved the prospects of long-term patient survival and thus prompted an increased concern with the durability of repair and with postsurgical appearance.

The profession's own historiography emphasizes the part played by war in advancing the discipline, arguing that because of changes in weaponry and casualty transport, each war, since the nineteenth century, has helped to hone different surgical skills. (For example, during the American Civil War, surgeons became skilled at treating disfiguring wounds of the jaw, a site particularly affected by ammunition introduced during that conflict.) In 1926, H. Lyons Hunt wrote about the effect of the First World War on the profession: "The ghastly brutality of the World War often left its victims maimed and disfigured and hence sensible that death would have been a kindlier reward for their patriotism. The demand for relief was imperative. Hospitals were established and surgeons who could not only manipulate a scalpel but could do so with consummate art were ordered to them. New methods and operations were devised, even surgical principles were changed or modified to meet unique conditions, and Plastic Surgery became a dignified speciality."[5] As Hunt indicates, progress during wartime was attributable not only to the increased number of injuries but to the confluence of talent found in the battlefield operating room.

Since its earliest days, plastic surgery has served a dual function: repair of deformity and cosmetic improvement (although there is no bright line to separate the two). Aesthetic surgery—defined as surgery performed to improve a normal, but less than optimal, appearance—began in the late nineteenth century, with operations to straighten hooked or crooked noses or to flatten protruding ears.[6] The first textbooks devoted to cosmetic surgery were published in the early twentieth century. The first face-lift has been the subject of a priority dispute for nearly one hundred years. Such an operation was described in an article published in 1912, but later, when challenged, the author claimed to have been performing the procedure as early as 1901.

Wartime improvements in surgical technique aided the progress of cosmetic work. Many surgeons initially trained in the repair of disfiguring war wounds returned to civilian life eager to use their new skills. In addition, as Hunt suggests in the following passage, publicity about the successful repair of wounded soldiers' disfigurements stirred civilian demand: "The war ended. The injured returned. The reconstructive work of the plastic surgeons was now on exhibition. When it became known that plastic surgery was successful in returning these men to their places in civilian life, a new call developed for this type of surgery."[7] Historian Elizabeth Haiken, however, has challenged the idea that the demand for cosmetic surgery was a by-product of the nobility of war. Rather, she argues that in the United States it was a function of broader social trends:

urbanization and immigration in the early years of the twentieth century; the consumer culture that developed in the context of postwar affluence in midcentury; and the popularization of psychology reflected in the construction of the "inferiority complex" and "self-esteem," notions that provided justification for surgery in the latter part of the century.[8]

Attempts to organize plastic surgery as a recognized surgical specialty were spearheaded by surgeons returning from the First World War. These young men came back from the battlefield to find no recognition for their professional identity in civilian medical institutions. The reputation of plastic surgery suffered from the taint of "commercializing charlatan[s]," poorly trained surgeons whose attempts at aesthetic surgery often had disastrous results.[9] The men who formed the first plastic surgery professional organizations took as their goals the establishment of a structure for the training and employment of plastic surgeons, the dissemination of technical information, and, perhaps most important, the regulation and legitimization of the profession.[10]

The first professional organization in the United States devoted to plastic and reconstructive surgery was founded in 1921 under the name the American Association of Oral Surgeons (reflecting a focus on the repair of facial injuries during the war). Members, who were required to possess both the M.D. and the D.D.S. degrees, were selected through nomination and election. Total membership was limited. The association called itself a "clinical society," describing its mission as "to stimulate and advance knowledge of the science and art of plastic surgery and thereby improve and elevate the standard of practice of this specialty."[11] In 1927 the association changed its name to the Association of Oral and Plastic Surgery.

In 1931, a group of surgeons who had been excluded from the association because they did not have training in dentistry formed their own organization—the Society of Plastic and Reconstructive Surgery. The constitution of the new society listed the following objectives: "1) to further medical and surgical research pertaining to the study and treatment of congenital and acquired deformities, 2) to keep the medical profession informed of the scientific progress and possibilities of plastic and reconstructive surgery, 3) to stress the great social, economic, and psychological importance of this surgical specialty."[12]

Founding members of the society chose an emblem that represented the dual nature of plastic surgery as art and science and reflected the profession's devotion to both beauty and healing: "The center figure is the Venus de Milo, Roman Goddess of Beauty. . . . In the lower foreground is the gloved hand of the creative surgeon-artist. The founding date, 1931, and these significant figures are encircled by a double Corona in which the name of the American Society of Plastic and Reconstructive Surgery is lettered and at the base is a horizontal Caduceus with the staff of Hermes and the Sacred Serpent of Aesculapius."[13]

Over the next ten years, through a series of changes in name and membership requirements,[14] the two organizations worked together to consolidate the status of the profession. In 1937, they established the American Board of Plastic Surgery, a governing body that would develop standards for training programs and guidelines for the certification of individual surgeons. A year later, the board was recognized as a subsidiary of the American Board of Surgery. In 1939, it was incorporated and issued its first list of approved residency programs. In 1941, the American Medical Association recognized the board as a major specialty board under the jurisdiction of the American Board of Medical Specialties, allowing it to certify surgeons meeting board criteria and to standardize requirements for residency training.[15]

The next fifty years were a period of steady growth for the profession. In 1946, the ASPRS had 100 members; by 1949, there were 150. Data are not available for the 1950s, but throughout the 1960s the total number of plastic surgeons grew steadily, from 1,023 in 1963 to 1503 in 1969. In the 1970s, the number topped two thousand, and in the 1980s, almost four thousand. By the 1990s, there were five thousand board-certified plastic surgeons in the United States.[16] This rate of growth was over twice that of the medical profession as a whole.[17]

The American Board of Plastic Surgery controlled the number of plastic surgeons by regulating the number of approved residency programs. In 1959–60, there were 69 hospital-based residency programs with 122 positions. By 1969–70, there were more than 250 positions at seventy-five hospitals. The large increase in the number of plastic surgeons in the 1980s was presaged in the 1970s by the establishment of 112 programs with 450 spaces. By the early

1990s, the number of programs had declined, down to 101, but the number of residency positions had decreased only slightly.[18]

Increases in the number of plastic surgeons were reflected in the increased number of surgeries taking place each year. For example, in the plastic surgery unit at St. Francis Hospital in San Francisco in 1959–60 there was an average daily census (ADC) of twelve inpatients and 360 outpatient visits. In 1969–70, the ADC was six, and there were more than 500 outpatient visits. In 1979–80, the ADC was thirty-four, and there were more than 8,500 outpatient visits. Similarly, at the Albany Medical Center Associated Hospitals in Albany, New York, over the same period of time, the numbers showed a similar pattern of increase, from an ADC of eighteen with 221 outpatient visits to an ADC of twenty-two and 6,000 outpatient visits.[19] "The past decade has seen a dramatic increase in the number of well-trained, talented plastic surgeons throughout the country," observers noted in 1978. "With expert surgery available in almost every community, public acceptance and demand for plastic and reconstructive procedures have rapidly increased."[20] The huge growth in outpatient visits and the relative stability of inpatient numbers suggest that the greatest increase was in the fairly simple, quick-recovery operations—many of which were cosmetic procedures.

The emphasis on cosmetic work was grounded in economics. In a 1955 article titled "The Crossroads of Cosmetic Surgery," a young plastic surgeon had argued for more attention to cosmetic procedures in residency training. His rationale was that for the vast majority of plastic surgeons in private practice, economic survival would depend not on the performance of reconstructive procedures—which were very costly and fairly rare—but on high volume and "high quality cosmetic work for people of all economic levels."[21] Sociologist Jane Sprague Zones described how this argument was soon institutionalized as a deliberate strategy. Following the explosion in the number of qualified practitioners, "rather than lowering the costs of professional services to increase their market, plastic surgeons . . . sought to increase demand for their services in order to maintain [their] income, which averaged . . . among the highest in medicine."[22] Key elements of this strategy included promoting plastic surgery to the public and making widespread access easier by establishing financing plans.

A 1988 report by the American Medical Association (based on data collected in 1986) provides a snapshot of the profession on the eve of the silicone breast implant controversy.[23] Plastic surgery was an overwhelmingly male specialty; 94 percent of its practitioners were men. It was also a young specialty; more than 55 percent of plastic surgeons were age forty-four or younger, while fewer than 5 percent were sixty-five or older. The five states with the greatest absolute number of plastic surgeons were (in descending order) California, New York, Florida, Texas, and Pennsylvania. On a per capita basis, however, there were more plastic surgeons in Washington, D.C., than anywhere else. (Iowa and Maine had the fewest.) Plastic surgeons represented less than 1 percent of all medical specialists. In 1988–89, ASPRS members performed 681,070 cosmetic procedures (an increase of 80% over the number in 1981) and 1.1 million reconstructive procedures (a 50% increase since 1981). Of those procedures, about 70,000 were breast augmentations and about 34,000 were breast reconstructions.[24]

The silicone gel breast implant played an important part in the economy of plastic surgery. Average charges for augmentation, which ran about $1,500 in the mid-1970s, increased to $2,000–3,000 in 1980, and to $4,000–5,000 in 1990. Breast reconstruction cost considerably more, about two to three times the price of augmentation, depending on the complexity of the procedure.[25] A 1992 article estimated that in the 1980s, at the height of their popularity, implants accounted for approximately one-fourth of the income generated by plastic surgeons—for a total of about $300 million each year.[26]

As plastic surgery has been professionalized, so too has its knowledge been rationalized. The knowledge(s) of plastic surgery include specific surgical methods and the materials and tools with which they are performed—for example, the technology and techniques of breast implantation described in chapters 2 and 3; skills in visualization and surgical planning; diagnosis, or the ability to apply a nosology of conditions to individual patients; patient relations and communication skills; and the evaluation of results. In addition, plastic surgery claims an overarching knowledge, a philosophy that interprets the significance of physical transformation.

From the earliest days of the profession, plastic surgeons have collected their knowledge in publications: newsletters, textbooks, journals, meeting and symposia transactions. This knowledge dissemination function was specified in the charters written at the founding of the professional societies and was soon a focus of the organizations' efforts. In 1940, the Society of Plastic and Reconstructive Surgery published a report of its annual meeting. *Plastic and Reconstructive Surgery*, the journal of the ASPRS, was founded in 1946 to publish articles on recent advances in surgical technique and to keep readers informed about the actions of the society. Initially issued six times a year, in 1950 it became a monthly publication.

Knowledge is disseminated on a daily basis through formal training and informal mentoring and, at regular intervals, through sponsored meetings. Such education was another early priority for the professional societies. In 1948, the ASPRS formed the Education Foundation, whose "mission was to support research pertaining to congenital and acquired deformities, promote high standards of training, practice and research in plastic surgery, confer scholarships and prizes, and promote lectures, seminars and medical and public meetings to educate the public in plastic surgery matters."[27] (That public education was part of this mission suggests that the profession knew that without demand there would be little to occupy its members.) The annual meeting of the society originated in the tradition of "stated meetings," where members with geographic proximity gathered to talk about cases and techniques. As membership grew, these meetings became more formal annual events.

The substance of this knowledge, and how it is composed, constitute the epistemology of plastic surgery. Several aspects of this epistemology are particularly important to understanding the history of breast implants and the FDA controversy: the progress of technical knowledge, the plastic surgeon–patient relationship, and how plastic surgeons understand and describe the philosophy of their specialty.

One historian has argued that technical progress in plastic surgery proceeds through two types of innovation: "firstly by the independent and original innovations of different surgeons in different parts of the world . . . and, secondly, because other surgeons have studied an earlier man's work and have then improved upon it."[28] That is, technical progress may be attributed both to true innovation and to modification. The former is much less common than the latter. In fact, modification seems to be the focus of the discipline's creativity: "Subtle modifications and alteration . . . [of standard, widely accepted, and time-proved procedures] are the hallmark of the creative surgeon."[29]

Technical innovation and modification are important for several reasons. They provide the means by which seemingly intractable problems are solved. (Such problems prompt innovation and modification.) By improving results, they make plastic surgery more attractive to a wider audience, thus broadening the patient base and providing more work for surgeons. For the individual surgeon, developing a popular innovation or modification promises a type of immortality (the early silicone implants, for example, were always called "the Cronin"), as well as the satisfaction attendant upon successful problem solving. Working against creativity, however, are other forces: "Aside from the usual inertia most of us have with respect to change, there is also a reluctance to try lest we fail. In addition to the distress to the patient, there might be the added consequence of litigation."[30] The competing pressures for and against technical progress cause plastic surgery to advance in the cyclical manner—from innovation to modification to disillusion—seen in the history of breast implant technology.

The organizational history of the plastic surgery profession is one of success in regulating the number and practices of its members. In that way, plastic surgery closely follows patterns set by other medical specialties.[31] Plastic surgery, however, differs from other specialties in the particulars of the plastic surgeon–patient relationship. In contrast to the professional detachment promoted by other specialties, plastic surgeons emphasize and encourage a psychological closeness between practitioner and patient, arguing that such closeness is essential for success.[32]

The plastic surgeon's emotional symbiosis with his patient begins with an ability to empathize with the patient's desire for physical transformation. In a world often hostile to individuals who seek cosmetic improvement through surgery, the plastic surgeon is able to remain nonjudgmental: "He does hear the appeal of the patient for the right to a place in society, for the removal of a barrier to his economic efficiency, for a share of happiness in his most intimate existence."[33] The plastic surgeon claims to know the patient as a "whole" person and to understand the links between physical appearance and psychological health: "When the entire problem of the patient has been revealed and is understood, often it becomes apparent that correction to normal of a physical irregularity . . . may be of utmost importance in the prevention of psychological pressures, which, in turn, through psychosomatic effect, may result in organic disease."[34] The empathic surgeon can intuit the patient's internal world: "It must be a chain of psychological reactions, unpleasant experiences, lingering remarks, disturbing self-consciousness which brings the patient to submit to a cosmetic operation. . . . [Patients] want to be inconspicuous, they want to rid themselves of (the feeling of being conspicuous by) an ugly unattractive feature."[35] So vital is the ability to empathize with the patient that "the importance of technical skill per se is diminished as one takes into account the urgency of understanding exactly what motivates the patient."[36]

But it is the plastic surgeon's professional responsibility not to cede important decisions to the patient. Technical questions of surgical technique, of course, always remain the surgeon's purview. As we saw in chapter 3, however, a few surgeons insisted that they, and not their patients, should choose the size of the breast implants to be inserted. At times, some plastic surgeons have suggested that this kind of control be extended to other aspects of the patient's life: for example, an overweight patient should be advised to lose weight; a middle-aged woman seeking a face-lift should be told to get a new hairdo as well.[37]

The dark side of all of this emotion is the potential for it, like a love affair, to turn sour. The necessary empathy may encourage an obsession on the part of the patient: "The understanding surgeon who grasps the specific emotional significance to the patient of having small breasts often will see the development of a unique patient-surgeon relationship and may even encounter problems with overly devoted patients."[38] Plastic surgeons attempt to limit the possibility of this eventuality, and other risks, by practicing "patient selection."

The earliest plastic surgeons recognized danger in certain patients: "Let the author warn the operator against the 'beauty cranks' especially of those who are just about to engage in great theatrical ventures, circus performers, or 'acts' and very desirable marriages. These are patients who are not only difficult to deal with, but the first to harm the hard-earned, well-deserved reputation of the surgeon and to drag him into courts for imbursement for all kinds of damages, especially backed up by events losses, and sufferings largely imaginable and untrue, and oftimes entirely impossible."[39]

A subtext of the "psychology of the flat-chested woman" was this issue of patient selection—determining just which patients might represent a risk of "harm" to the surgeon.[40] While the "supermarket attitude" of patients—the "feeling that their ability to pay should qualify them for the operation they desire"[41]—was troublesome, the greater danger lay in the possibility of a specific psychopathology. This psychopathology was conceptualized as "somatic delusion," a condition characterized by a skewed reality in which "an insignificant deformity appears conspicuous, attractive features appear ugly." Sufferers demonstrated "a projection of dissatisfaction in other areas onto [the] body part; concretization, [in which] one area becomes [the] focus of all body ills; [and a] preoccupation with [the] body part to the extent of taking inappropriate actions."[42] Such patients tended to be "chronically dissatisfied" with the results of surgery. They "return again and again because of insignificant imperfections that have become invested with inordinate emotional importance. After several unsuccessful corrections, the surgeon realizes that these patients can never be satisfied, because he himself has become—in the patient's mind—responsible for the original defect they were seeking to correct."[43]

Psychiatrists helped plastic surgeons to identify the warning signs of possible trouble in women seeking augmentation mammaplasty. Such signs included indications that the patient wanted the surgery to please someone else; a sense of urgency; unrealistic or vague expectations; a competitive component to the desire for surgery; no objective evidence for the deformity; a request for surgery during or after a major life crisis; and postsurgery elation disproportionate to the change effected.[44] Some of these warning signs, of course, differed only in degree from what was considered "normal" in women seeking augmentation.

Women's magazines addressed the issue of patient selection in their discussions of the "good candidate." "In simplest terms," one article explained, "candidacy requires only that you are a woman and that you feel that your bustline is too small. But being a candidate does not guarantee that a plastic surgeon will agree to do the surgery."[45] The good candidate had to demonstrate that she sought breast augmentation "for herself," not to please others or in order to "change her life."[46] (Actress Mariel Hemingway, who in the early 1980s made a highly publicized decision to enlarge her breasts in order to win a movie role as a Playboy bunny, told *People* magazine that she had done it "for herself."[47] In an article about the new voluptuous style and how fashion models were seeking augmentation surgery at an increasing rate, plastic surgeon Steven Herman told a reporter he believed "that most shell out the $5000–6000 fees for their self-image rather than out of a belief that an increase in bust size will mean an increase in bookings.")[48] Over and over again, readers were warned that surgery could not save a marriage (women who thought it would do so "need a psychiatrist or a marriage counselor, not a plastic surgeon")[49] or treat "deep seated psychological problems."[50] Yet the same magazines reported that augmentation surgery could "provide astonishing psychological relief."[51] One author asked, "Who is a good candidate for cosmetic surgery?" and answered by explaining that the good candidate had realistic expectations and clearly defined goals, was committed to change, engaged in open communication with her surgeon, was in touch with her own feelings, and was "appropriately" motivated.[52] (By contrast, popular articles about reconstructive breast surgery included little discussion of the notion of candidacy.)

At the height of the controversy over silicone gel breast implants, the ASPRS published a primer on malpractice.[53] Written by a plastic surgeon, it examined the causes of patient-filed malpractice claims, arguing that "well over half the claims are preventable. Most are based not on technical faults, but rather on failures of communication and patient selection criteria"—failures that were best avoided by "be[ing] honest, warm and compassionate . . . [by] behav[ing] toward the patient as you would want another physician to behave toward your spouse," and by carefully assessing the patient's motivation.

The "root cause" of litigiousness was misplaced patient anger. "The borderline between anxiety and anger is very tenuous, and the conversion factor is uncertainty—the fear of the unknown. How do we cope with uncertainty? Blaming someone else places the responsibility elsewhere and gives one a sense of 'control' which, however inappropriate, is easier to cope with psychologically." Insight into this dynamic might allow the surgeon to head off litigation: "The perception of the patient that you understand the uncertainty and will join with him or her to help conquer it may be the deciding factor in whether that patient's next move is to seek legal counsel. . . . the technique of attentive silence often defuses angry people. . . . One of the worst errors . . . is to try to avoid them. . . . the more you talk and listen to that patient, the more likely you are to avoid converting an incident into a claim."

Other plastic surgeons contended that there were limits to the power of communication: in particular, that women did not hear what they were told about the risks of surgery and the possible complications of implants. In one study, patients were reported to have "demonstrated marked repression and denial of any serious complications; they could not recall these risks."[54]

The significance of the plastic surgeon–patient relationship lies in its impact on the legitimacy of the profession. Paul Starr argued that "the ideal of a profession calls for the sovereignty of its members. . . . A professional who yields too much to the demands of clients violates an essential article of the professional code: Quacks are . . . practitioners who continue to please their customers but not their colleagues."[55] The struggle of plastic surgeons to establish their legitimacy was a struggle to overcome this definition of quackery. The oversight and control of training instituted during the middle part of the twentieth century had set high standards for the technical skills of plastic surgeons. It was the demand-driven nature of cosmetic work that remained questionable.

Cosmetic surgeons bemoaned their ostracism from the larger medical community. In 1957, one surgeon noted that "many people are still under the impression that cosmetic surgery is a nonserious, frivolous member of the somber plastic family."[56] Fourteen years later, little had changed. "Plastic surgery has long been established and fully recognized as a specialty," wrote one observer. "Yet . . . [cosmetic surgery] is still looked upon with criticism, skepticism, and even disdain by fellow physicians."[57] This disdain was due both to the perception that cosmetic procedures catered to empty vanity, and to prejudice against the population seeking aesthetic surgery, a population stereotyped as rich and idle. Cosmetic surgeons could be viewed as little more than workers for hire to the rich. In a specialty where patient demand was primary, such practitioners were subject to patient whim rather than to the demands of their professional colleagues.

Cosmetic surgeons' perception of their own professional ostracism was echoed in their descriptions of societal objections to patients who sought breast implants—objections to vanity or frivolity which they characterized as based on ignorance and prejudice. In emphasizing their joint ostracism, plastic surgeons reinforced the emotional symbiosis with their patients. In a cyclical manner, then, the close surgeon-patient relationship became both a cause and an effect of the extant professional and social prejudices, whether perceived or actual.

In the foreword to a 1984 volume on breast reconstruction, a plastic surgeon wrote: "The art and science of plastic surgery . . . comprise the study of *choice* and *proportion*. We do not or should not learn and teach *techniques and how but a philosophy and why*."[58] This philosophy situated the need for plastic surgery in a medical context and a social-historical-cultural context of the human desire for decorative physical adornment.[59] The same art/science duality is reflected in discussions of diagnosis. Plastic surgeons treat both disfigurement and variation perceived to be unaesthetic. The word "deformity" carries both connotations. That is, it may refer to bodily malformations that affect the individual's daily functioning, or it may connote an appearance that is less than optimal.

Introspective surgeons, reflecting on the dual nature of the specialty, have suggested a tension between plastic surgery as medical science and plastic surgery as art—"the specialty nearest sculpture in the living."[60]

> Art is the sublime manifestation of man—his share of divinity. Matter and shape stimulate the soul of the artist to generate beauty. This beauty is molded partly through the ability of his hands but is mostly derived from the refinement of his feelings. Surgery represents a form of expression akin to sculpture but the list of our Michelangelos will always be limited, for the range of a surgeon's artistic spontaneity is always restrained by the rigid laws governing surgical techniques, the respect that must be maintained for human tissue, and the depth of understanding required of our daily models, the patients. We need not apologize for our short list of masterpieces, for the palette and brush we offer in the specialty of plastic surgery are creativity itself and the very generous dividend of a satisfied patient.[61]

The tension between art and science is partially resolved in explanations of the goals and effects of plastic surgery, which are portrayed as collaborations between surgeon and patient. The surgeon brings the skill, judgment, and experience. The patient furnishes the body, the raw material. They both desire transformation. Success or failure is to be judged by both parties. Although plastic surgeons distinguish between objective and subjective results, the objective assessment is not always the surgeon's. In fact, as noted earlier, the surgeon is described as having to "live with" the results of his own work. He makes subjective assessments of surgical success based on (subjective) aesthetics and the effect (on his own life) of having performed a particular surgery. Surgical success is also dependent on the emotional transference between surgeon and patient. (See chapter 4.) This transference resides in a psychosexual realm where surgeons are father/lover figures and (male) surgeons manage empathy by equating the breast with a penis.

Plastic surgeons believe that they can change the inner person by changing the exterior: "We have always known that inner psychic and spiritual changes bring about a new external radiance, but we are now discovering that the process also works in reverse: Change the external appearance . . . of a person struggling continually against indifferent or negative social relations, and the inner light that has died within begins to glow once more," one surgeon wrote.[62] Another argued that aesthetic surgery "is both science and art. It is science insofar as it demands the most skillful employment of the subtle and inconspicuous surgical modalities in the attainment of its objective. It is art insofar as its objective is recognized to be pleasure for the patient."[63] Thus the art and science of plastic surgery are reconciled by the ultimate goal: happiness as represented by patient satisfaction.

The history of plastic surgery created a specialty that, while mainstream in its technical application, contained distinct elements in theory and in practice. These characteristics grew out of an emphasis on the subjectivities of both surgeons and patients. That is, every aspect of the plastic surgical encounter, from the designation of need to the assessment of results, was highly individualized, grounded in the personal and the emotional. As silicone breast implants became the subject of a public controversy and plastic surgeons found themselves important advocates for the continued availability of the devices, this subjective, personal approach was reflected in their strategies and claims.[64]

The FDA advisory panel that in 1982 recommended that implants be placed in Class II—a classification that would not have required any further study of safety and effectiveness—was composed largely of plastic surgeons. (As Silas Braley had predicted, the FDA had turned to the plastic surgeons as experts.) The members of the panel concluded that implants had "demonstrated a reasonably satisfactory level of performance over a long period of time," basing this conclusion on their own "personal knowledge of, and clinical experience with, the device." Their reading of the scientific literature reinforced this conclusion: although they noted that there had been several reports of adverse effects, they interpreted these as anomalies, the result of "errors in surgical technique."[65]

The panel's recommendation was consistent with the epistemology of plastic surgery in that it granted validity to personal experience—in this case, the plastic surgeons' own clinical experience—as evidence. It was also consistent with history. As with paraffin, the plastic surgeons were quick to blame individual susceptibility, then failures of surgical technique, and, only last, the material itself. In both cases, these hierarchies of blame reflected a subjectivist focus that looked first to individuals. To the plastic surgeons, even evidence that seemed to implicate the material (paraffin or silicone) did not suggest a need for outside intervention.

Rather, it was an opportunity for modification and, eventually, innovation.

The FDA disagreed, however, announcing that it was rejecting the panel's recommendation and putting implants in a stricter regulatory category. The agency's notice of intent to place implants in Class III prompted a swift response from the ASPRS. In a document submitted as public comment to the agency's official announcement, the society argued that such a classification was "inappropriate and unnecessary." Its evidence for this assertion was, once again, grounded in the particular epistemology of the discipline.

The document pointed to the innovation-modification-disillusion cycle as the profession's own form of self-regulation, suggesting that if left alone plastic surgeons could solve the remaining problems with implants. It repeated the defensive pattern observed some seventy-five years earlier with paraffin injections, blaming individual susceptibility and surgical technique for adverse effects. It portrayed the profession's definition of "deformity" as both functional and aesthetic. It equated effectiveness with psychological benefit and psychological benefit with patient satisfaction, implicitly asserting that plastic surgeons were the only ones with expertise in providing and understanding such benefit. It argued that complications were also subjective, the result not of inherent problems with implants but of failure in patient selection. Finally, the document reflected the profession's sensitivity to the charge of frivolity.

In a passage that during the controversy was widely quoted ("mocked" might be a better word) as an example of how financial interests had clouded the profession's judgment, the ASPRS claimed that implants were the treatment for a disease.

> There is a common misconception that the enlargement of the female breast is not necessary for the maintenance of health or treatment of disease. There is a substantial and enlarging body of medical information and opinion, however, to the effect that these deformities [the female breast that does not achieve normal or adequate development . . . mastectomy for cancer] are really a disease which in most patients results in feelings of inadequacy, lack of self-confidence, distortion of body image and total lack of well-being due to a lack of self-perceived femininity. The enlargement of the underdeveloped female breast is, therefore, often

> very necessary to insure an improved quality of life for the patient.[66]

Critics leaped, in particular, on the word "deformities." But as we have seen, for the plastic surgeons deformity was a continuum that could include unaesthetic variation in cases where it caused unhappiness.

The emphasis on benefit harked back to the 1950s, and the efforts of plastic surgeons to establish a benefit to balance the risks of sponge implants, and to the 1970s, when, with the introduction of the apparently "ideal" silicone technology plastic surgeons had expanded the definition of need, revising the notion that the necessary indication for implant use was a kind of pathology. Now that revision—and the old charge of frivolity—were both coming back to haunt the profession. The FDA had ignored benefit, the ASPRS argued, "impl[ying] that it perceives the benefits of breast implants to be slight or inconsequential, or 'merely aesthetic.' In fact, breast implants confer substantial benefits to patients with direct and valuable contributions to their quality of life, well-being, and health. In the context of these benefits . . . the record of safe use . . . overwhelmingly justifies a finding that premarket approval is not required." That "record of safe use" was a tribute to the profession's success in regulating technology itself through the innovation-modification-disillusion cycle: "Improvements in breast implants . . . as well as improved surgical techniques have resulted in minimal complications with the use of these devices in recent years."

The ASPRS challenged the FDA's list of complications and potential risks. This challenge was based on different interpretations of evidence and different conceptualizations of risk. First, the society argued that the risks were not significant: "The incidence and severity of adverse effects associated with breast implants are not significant"; then, that these insignificant risks really had nothing to do with the device itself (and thus could not be alleviated through more stringent regulation). "The majority of the complications identified by the FDA . . . are, in fact, not device-related complications. Rather, they are the natural and expected consequences of surgery or the result of physician error." In fact, these complications were not complications at all. Capsular contracture, for example, was "not a complication, but rather a normal event in *all* wound healing."

By asserting that a "phenomenon cannot be considered a complication if it does not produce patient dissatisfaction," the document suggested a new definition of complication (and a new way of assessing risk), one grounded in the subjectivist epistemology of the profession. The society then went on to dispute the validity of consumer complaints to the FDA, arguing that these "dissatisfactions are not . . . typically the results of surgery per se, or any complications of surgery, but may be a function of poor patient selection." By invoking patient selection, the plastic surgeons revealed one of its implicit purposes: to grant validity only to certain subjects.

The ASPRS looked at the problem of implants as one not of health and safety, but of misunderstanding and prejudice. Both implants and the profession were misunderstood—maligned and denigrated by societal insensitivity and ignorance. If the problem was one of misunderstanding, information could be the solution. Thus the ASPRS moved to control the production of information about implants.

The FDA's 1982 decision to place implants into the regulatory category that required premarket approval of safety and effectiveness resulted in a struggle to define these terms and prove how implants met (or did not meet) the evidentiary criteria. Safety and effectiveness were not completely foreign concepts to the plastic surgeons, but the meanings they gave the terms and their procedures for establishing them revealed a chasm of difference between the profession and the regulatory community.

During the shift to synthetic materials in the 1950s, the attention to "ideal properties" was a way of posing questions about the safety of implant materials. The early animal studies certainly fell within the boundaries of traditional scientific assessment. Divergence from this kind of assessment emerged as implants came into clinical use. Plastic surgeons tended to work alone or in small group practices. While some had ties to academic medicine through their reconstructive work or as residency preceptors, the academy had shown little interest in investigating breast implants. There was no good source of data about the number of implants in use.

Follow-up times tended to be short. The grounding of the surgeon-patient relationship in subjective assessment of aesthetics gave it a different flavor than the traditional objective and authoritarian relationship assumed in research. For all of these reasons, no long-term assessment and nothing resembling a controlled trial—the FDA's sine qua non for proving safety and effectiveness—were ever performed.

As it became clear that the FDA would insist upon such a demonstration, plastic surgeons struggled to elucidate the epistemological differences between the agency's definitions of safety and effectiveness and those of the profession. The discordances between the FDA and the plastic surgeons were illustrated vividly in a presentation made at the 1983 Annual Meeting of the ASPRS by plastic surgeons Mary McGrath and Boyd Burkhardt (and published the next year as an article in *Plastic and Reconstructive Surgery*).[67] An obvious response to the FDA's recent announcement of intent to classify implants in Class III, the presentation reviewed the work that plastic surgeons had done to evaluate implants and argued that there was no evidence to support the agency's contention that implants represented a potential risk.

"Implicit in the recent reservations about implants is the suggestion that plastic surgery has failed to promote scientific inquiry in this area," the authors began. In fact, they argued, there was no dearth of research on breast implants: "251 papers on the use of breast implants for augmentation mammaplasty have been published in the last 20 years."[68] Many of these papers were descriptions of the different types of implants available and the different surgical techniques for implantation. Others examined complications, including ninety nine papers on capsular contracture. The authors argued that such a volume of literature evidenced the profession's interest in and responsiveness to questions of scientific investigation.

Their definition of safety, however, differed from that of the FDA. The plastic surgeons' definition was bounded by clinical experience. (In contrast, the authors implied, the FDA's was based on innuendo.) Plastic surgeons had paid little heed to certain safety concerns—for example, breast cancer—because "the composite clinical experience has not mandated it."[69] Their frustration resulted from the fact that the FDA was arguing that "the mere absence of such reports is not valid scientific evidence that these risks do not exist."[70]

Efficacy, McGrath and Burkhardt wrote, "is defined as the probable benefit to health weighed against any probable illness or injury from the use of the device. Organizational guidelines for the clinical studies needed to document efficacy include providing confirmation of the medical condition being diagnosed and treated, assigning patients to test groups with controls and 'blinding,' and quantitative evaluation of the results in the treatment group compared with the control group. In other words, the scientific method."[71] Unfortunately, the authors argued, the particular circumstances of cosmetic breast surgery made it incompatible with such rigorous assessment: "Documenting the patient's preoperative pathology is subjective, comparison with untreated controls is unrealistic, and evaluation of the benefit to the patient truly involves only her own satisfaction with her self-image postoperatively. . . . It is the patient's subjective response . . . that is the determinant of benefit, or efficacy."[72]

The plastic surgeons and the FDA might as well have been speaking different languages. Plastic surgeons saw clinical experience and scientific evidence as equally valid, refusing to privilege the latter over the former, as the FDA required. Because of their equation of experience with evidence, the plastic surgeons rejected any suggestion that silicone was unsafe. Thus their focus shifted to effectiveness, which they called "efficacy" and construed as entirely subjective.[73] The FDA's standards for measuring effectiveness were, they argued, incompatible with the kind of benefit offered by implants. When the terms of the assessment were redefined in these ways, the balance clearly favored implants.

Although it was 1982 when the FDA announced its intent to place implants in Class III, the agency made no further move to do so until 1988. After that long delay, there were several more lulls as months passed to allow time for public comment (a statutory requirement) and, then, for the manufacturers to prepare their applications (another requirement). By 1990, however, the issue was attracting sustained attention.

When in December of 1990 the ASPRS was denied permission to testify at the Weiss hearing, it contributed a written statement that was later published as part of the hearing record.[74] In the statement the society objected to its exclusion, noting that while it had been justified by the committee's ostensible focus on regulatory issues, not implant safety, the safety issues had in fact been paramount. The one-sidedness of the presentations had vilified plastic surgeons.

The ASPRS argued that the testimony presented at the hearing had been "misleading and inaccurate." "Current evidence *does not warrant removing any type of breast implant from the market at this time*," the document stated, noting a lack of evidence linking implants to cancer or to autoimmune disease. About the latter, the plastic surgeons reported that they had been advised "by many epidemiologists and other experts" that "due to the rarity of the disease, conclusive data could not be obtained." The statement also emphasized that the ASPRS had been proactive in soliciting research on the possible adverse effects of implants and had worked hard to make the latest data available to its members and their patients—even before "the relatively recent interest of Congress, FDA and various consumer groups." Such attempts of gathering and disseminating information were examples of the society's "history of concern for patient safety and our desire for an open exchange of information." In closing, the statement called for the presentation of "more balanced information": "We, like you, are primarily interested in patient well-being, and ask only for a balanced risk-benefit assessment."

The ASPRS statement framed the problem of implants as one of misunderstanding that encompassed both implants and the profession itself. This misunderstanding had been promoted by various interest groups. By submitting its statement, the society was attempting to correct these misunderstandings by arguing for the safety of implants and by positioning plastic surgeons as concerned scientists with a history of conducting valid research on implants.

The theme of the misunderstood implant was prominent in "Straight Talk . . . about Breast Implants," a patient education brochure produced by the ASPRS in 1990.[75] The brochure declared that "breast implants are among the safest surgically implanted devices in use today," but its subtle messages are more interesting—in particular, its portrayal of the surgeon-patient relationship and its stunning reconceptualization of the adverse effects of implants.

"A woman's desire for breast implants is often misunderstood by friends and relatives, who may wonder why anyone would risk pain and potential complications for what they perceive as frivolous vanity," the document stated. Plastic surgeons, however, understood that "the desire to improve one's appearance is universal to *all* human beings; we vary only in the extent to which we will go to achieve our fullest potential." Because of this empathy and their technical knowledge, plastic surgeons were "the best source of [information about implants] . . . since the 'facts' on breast implants are often incompletely understood by the public and the popular press."

Although the brochure admitted that implants posed potential risks, it reconceptualized these risks to liken them to the natural. It argued that "for the vast majority of women, the implants will come to feel like part of their bodies, occasionally subject to their own special 'illnesses' and injuries—just like a natural organ." It was the occurrence of these "special illnesses and injuries" that proved the naturalness of the devices. "Just as the kidney, heart, eyes or any other body part can fail, so can man-made implants."

"Straight Talk" portrayed the plastic surgeon as knowledgeable and benevolent, possessing an empathy superior to that of the woman's own family and friends. This construction reinforced the image of the surgeon-patient relationship extant in other sources: the surgeon and the patient are allied by desire and skill and brought closer together by the misunderstanding of others.

As the controversy drew more media, and thus public, attention, the plastic surgeons, as represented by the ASPRS, swung into action with an extensive media campaign. The society justified the need for an organized response by claiming that "biased and inaccurate media coverage" and a "political and regulatory climate" influenced by the likes of Congressman Weiss and Public Citizen's Sidney Wolfe were threatening the well-being of patients and the autonomy of the profession. [76] The media campaign sought to sway public opinion, and the opinions of legislators, by presenting a more "balanced account" of implants. The campaign, which the plastic surgeons always characterized as an "educational" effort, implemented both defensive and offensive strategies.

The defensive strategy began with the claim that scientific evidence and a long history of use proved the safety of implants. The "reliable, scientific evidence collected to date indicates that the majority of patients are not exposed to any significant health risks," the organization wrote in one press release.[77] Another ASPRS document urged surgeons to emphasize the "30 years' of implant usage and scientific research" in their letters to Congress and the FDA.[78] The society cited a 1991 poll in arguing that more than 90 percent of the estimated two million women with implants were satisfied with the devices, and blamed "a small number of anecdotal, sensational stories" on dissatisfied patients who had achieved undue political influence. [79] The ASPRS argued that the surgeon's job was to make known the views of the "silent majority."[80]

Embedded in the claims about the satisfaction of 90 percent of the women with implants, however, was the recognition that a minority might be suffering because of implants. The plastic surgeons supported more research about these cases, but their ready claim was to blame any adverse effects on individual susceptibility and point, again, to the majority benefit: "We believe breast implants are similar to birth-control pills and other devices or drugs. Just because they can cause ill effects in certain individuals does not mean everyone should be deprived of their benefits."[81]

The ASPRS also began to answer criticism about plastic surgeons' own role in the controversy. "Media Tips" in the *Breast Implant Bulletin* urged them to respond to "aggressive" questions about the ASPRS sponsorship of the October Fly-in (the 1991 event in which plastic surgeons and ASPRS-sponsored patients flew to Washington to lobby legislators) by "remind[ing] them that many of these patients are working women and mothers who did not have the resources to finance such a trip. The Society believed that the women themselves could tell their story most eloquently." To "the charge of financial self-interest," the society advised, "don't pull any punches. Simply point out that your income would be threatened much more if you had a lot of unhappy patients who sued you because they weren't happy with their results. Performing a procedure that isn't safe or effective just to get the money is not good business—or good medicine."[82]

The more vigorous strategy, however, was to take the offense. The offensive campaign had two parts: claims about the negative consequences of restricting implant availability and a rhetorical appeal to individual rights. If the FDA were to ban or restrict implant use, the ASPRS argued, the "need and desire for implants would not go away."[83] Rather, there would be dire consequences: Women would be dissuaded from seeking early detection of or treatment for breast cancer. Insurance companies would deny coverage for any breast condition in all women with implants. Women who already had the devices would be unable to get replacements in cases of implant failure. Manufacturers in the United States would be driven out of business as liability claims against them soared. Research and development of new medical devices would be discouraged. Unscrupulous operators and unreliable overseas manufacturers would rush into the void, servicing an entirely unregulated, unsafe black market in implants. Rich women would seek the procedure in countries where it was still available. FDA restrictions on silicone breast implants would threaten all medical devices using silicone, a number of which were lifesaving.[84]

These claims were directed at a variety of audiences. The ASPRS used them in internal communications to its membership, in letters that urged plastic surgeons to write to Congress and the FDA or to contribute money to the cause. The same claims appeared in "sample letters" that the ASPRS urged surgeons to send to their patients, so that they, too, could write to Washington. Finally, these claims were emphasized in official testimony and in media statements directed at both decision makers and the general public.

The appeal to individual rights appropriated a rhetoric of choice and personal freedom associated with the feminist movement for reproductive rights: "Each woman should have the right to decide" and "The right to make this decision is important to the women of this country" were both "key message points," suggested by the ASPRS for use in letters to Congress. The same language appeared in ASPRS press releases as quotations from women: "We want the right to decide for ourselves whether breast implants are right for us—*after* being fully informed of the known and suspected potential risks."[85] The rights argument was used to recruit both surgeons, who were told that they stood as their patients' advocates against the threat to personal liberty, and patients, for whom the threat was made personal and immediate: "Other women will not have access to the option you had for improving your self image. . . .You could be discriminated against and become 'uninsurable.' . . . If you were to need an implant replaced for whatever reason, it would be virtually impossible to have the surgery in the United States. . . . It is *imperative*—to protect your rights as well as those of millions of other women—that you act now."[86]

Recruiting patients as implant advocates was a crucial part of the ASPRS strategy. As the society explained to its membership (in a letter urging them to solicit patients to write letters to Washington), "patients are our best allies on this issue. Once they understand the biased and inaccurate nature of the current media coverage and the impact it may have on them, it's been our experience that they are willing to engage in letter writing campaigns as well. As the most affected parties, their letters have even more impact than ours."[87] In light of this deliberate recruitment, the ASPRS's attempt to position its members as "patient advocates" appears highly ironic.

The ASPRS "advocacy ad" that appeared in major newspapers in October of 1991 attempted to make use of that "impact." The ad (its visual impact was described in the introduction) read:

> During the past 30 years, more than two million women have had breast implant surgery. It has helped them overcome personal and psychological pain and improved the quality of their lives.
>
> Today, the federal government is threatening to take this option away.
>
> The Food and Drug Administration is considering banning or limiting the use of breast implants, even though it has not found them to be unsafe or ineffective. Indeed, scientific studies conducted during the past three decades failed to show a link between breast implants and harmful, long-term health problems.
>
> Plastic surgeons vigorously support continued scientific research on breast implants. They believe women must be fully informed of any risk and benefits. Most importantly, they support a woman's right to make her own, informed medical decisions.

> An ill-conceived government ban would have serious consequences. Women who desire implants—for any reason—to improve their psychological health and well-being could not get them in the U.S. Those who already have implants may not have access to replacements if needed. Other women may be denied insurance coverage for breast surgery or lose the chance for reconstruction after cancer.
>
> Severely restricting the use of implants would have the same effects. With only a few women eligible to have breast implant surgery, makers of implants could simply not afford to make them available.
>
> You can make a difference. You can help all women retain the right to decide for themselves about breast implants . . .
>
> After all, it's one of the most personal decisions a woman can make. Shouldn't it be her own?

This advertisement, which came under heavy criticism by groups of implant opponents,[88] hit all the notes of the ASPRS's refrain. It portrayed women as the primary advocates for implants. It emphasized reconstruction, implying (by the women pictured and quoted) that at least half of implant surgeries were for that purpose. It portrayed implants as an intervention that could effectively improve self-esteem. It repeated the warnings of dire consequences. By blaming the situation on intrusive federal government, it attempted to shift the burden of proof to the FDA, implying that the agency had failed in its duty to prove implants unsafe and ineffective and was acting chiefly out of malice. Finally, it appealed to an American ideal of individual rights and responsibilities to resolve the controversy.

The defensive claims made by the ASPRS portrayed implants as a problem being promoted by organized groups interested in achieving their own political agendas. The groups of implant victims—for example, organizations like CTN—were understood as conglomerations of dissatisfied patients intent on suing. The plastic surgeons responded to their claims as they would to similar claims from individuals. That is, they interpreted every claim as the complaint of a dissatisfied patient—the result of improper patient selection. Because they saw these patients as litigation risks, and because they understood that litigation was largely a problem of miscommunication, the profession's response was to focus on the emotional content of the victims' claims—their "hysteria"—rather than to assess the extent to which their injuries may or may not have been linked to silicone.[89]

The plastic surgeons' offensive claims served to reconstruct the problem of implants. While the FDA and the implant opponents sought to promote the problem of implants as one of risk or harm, respectively, the plastic surgeons suggested that the real problems were fairness and the threat to individual rights. "Need and desire" were unassailable, unquestionable because the FDA lacked both the statutory authority and the empathy needed to do so. By contrast, the plastic surgeons understood and stood as guardians against the authoritarianism of the state.

In their statements to the FDA advisory panel at the November 1991 meeting, the plastic surgeons introduced several new arguments. These new arguments had two objectives: to minimize the risks of implants by situating them in a world of risk; and to forestall a policy recommendation that might draw a distinction between reconstruction and augmentation. Several plastic surgeons noted that other products known to be harmful were available to adults on the open market. For example, the government allowed access to cigarettes,[90] alcohol, and over-the-counter drugs. By focusing on implants, the surgeons argued, the government "certainly provides a mixed message to patients. It condones or at least passively approves the use of tobacco and alcohol with minimal warnings and then we are banning a device that is clearly beneficial to the vast majority of cases based on known inherent risk and otherwise theoretical risk."[91]

That the FDA might draw a policy distinction between reconstruction and augmentation patients had been hinted at since the late 1980s. Such a distinction was implicit in the agency's emphasis on proving benefit and in its expressed concern that young women were the recipients of most of the devices. Preventing a distinction had been the subtext of many of the ASPRS's campaign claims. For example, it was apparent in the emphasis of the advocacy ad on the psychological value of augmentation, and it drove the organization's assertion that a restriction would do as much damage as a ban.

The strategy for preventing a policy distinction was to emphasize the value of implants for reconstruction patients, then challenge any attempt to distinguish between reconstruction and augmentation patients. The preponderance of the evidence presented by the plastic surgeons and their panels of experts dealt with reconstruction, thus creating the impression that implants were used primarily for reconstruction. The surgeons showed before-and-after photographs of congenital deformities and mastectomy sites that had been repaired with implants. In lists of indications, augmentation was always cited last. Discussions of psychological benefit emphasized the good done for reconstruction patients. When augmentation was mentioned, it was to draw an emotional or psychological equivalence between it and reconstruction. Because of this equivalence, to make a policy distinction between the two procedures would be "judgmental." As one female plastic surgeon told the advisory panel: "The opinion that women who have a medical reason to choose breast implants, whereas those of us who chose them for other reasons, psychological and emotional, are somehow unworthy, I find both judgmental and irrelevant. If implants are safe for the reconstruction patient, they are safe for all of us no matter what our motivation."[92] An ASPRS petition submitted to the FDA just after the November hearing argued that the "benefits derived from breast augmentation for emotional well-being are similar or identical to benefits derived from breast reconstruction."[93]

The petition also examined the risks cited by the FDA, categorizing them as "known and manageable" (e.g., capsular contracture, tumor detection, infection, calcification, and gel leakage and migration) and "speculative" (e.g., carcinogenicity and autoimmune disease). After its own review of the scientific evidence, the society concluded that "a serious consideration of the benefits and known risks of silicone gel–filled breast prostheses clearly shows that the continued availability of the device for fully informed women is unquestionably beneficial in that each woman will be able to exercise choice and determine for herself whether the benefits of the device outweigh its known and speculative risks."[94]

These new arguments appeared to give up some ground to the opponents of implants. But even while acquiescing to—at least the possibility of—risk, however, the plastic surgeons sought to control the situation by offering a solution to the problem that was consistent with their own epistemology. The solution was called "informed choice," with the emphasis on the subjective, individualized notion of choice.

On January 6, 1992, David Kessler rejected his advisory panel's recommendation and, based on "new information," called for a moratorium on the sale and distribution of silicone gel breast implants. Because the agency had no authority to regulate physician practice, it could only request that the plastic surgeons cooperate. The moratorium crystallized the ASPRS's grievances against the FDA. Specifically, the ASPRS accused the agency of bias in its dealings with the plastic surgeons. The FDA had refused to acknowledge or respond to the society's communications—the various petitions, and so forth—and had treated the ASPRS as an adversary. It was dictating the studies that should be done without any input from the plastic surgical community. The agency was unaware of the best data available—data resulting from the plastic surgeons' own experience. The advisory panel, now almost devoid of plastic surgeons as voting members, was unqualified and biased. "Our observation of the entire FDA process leaves us with an empty feeling, the impression that conclusions are already established before all the evidence is considered."[95] Once again, it seemed, plastic surgeons were getting no respect.

During the February 1992 hearing, as explored in chapter 5, those fighting for the continued availability of implants made the charge that "pathologic science" was responsible for the FDA's actions. There was an irony in the plastic surgeons presenting themselves as exemplars of scientific objectivity, but more significant was their apparent realization that the ground had shifted—that empathy would not be enough to prove safety and effectiveness.

In the wake of that realization, the plastic surgeons unleashed a new defense: a seemingly contradictory strategy that both praised and blamed members of the profession. One surgeon argued that implant surgery was the "obstetrics of plastic surgery": "I think most obstetricians would prefer to be gynecologists and not get up in the middle of the night and get paid better, but there's great joy and satisfaction in the presentation of new life. . . . Our motivation comes from our patients. We don't go out and seek them. They seek us out, and the satisfaction that we perceive, the quality of life improvement that is repeatedly told to us, day after day after day in the consulting room, for many, many years after they've had the surgery, is the reward and the motivation."[96]

Juxtaposed against this altruistic portrayal of the profession, however, were other statements implying that the problems with implants were the result of incompetent and dishonest plastic surgeons. One surgeon suggested that the high rate of implant rupture was an artifact, the result of plastic surgeons lying to manufacturers about their own surgical errors in order to get refunds for the damaged devices.[97] Another acknowledged the possibility that some surgeons had failed to inform their patients about the risks of implants.[98] These were individual actions, however, not indications of implant risk. Such actions were outside the jurisdiction of the FDA, and would be dealt with within the profession. In a sense, by adopting this version of the problem, the plastic surgeons were reviving an old problem—one that the profession had successfully managed before—that of the commercializing charlatan.

But this final attempt to divert attention from implants was unsuccessful. When the panel recommended restrictions on the availability of implants, there was little opposition from the plastic surgeons. The ASPRS called the panel's recommendation "discriminatory" but seemed to recognize the inevitability of the FDA's final policy decision. For the time being, the problem of implants was reconstructed as the damage that the controversy had done to the profession. The focus shifted to repairing that damage.

The strategies used by the plastic surgeons during the controversy become more meaningful—and understandable—in the context of the history of the profession. The discipline's conceptualization of itself as a manifestation of art and science united in the goal of individual happiness led the plastic surgeons to equate their own attempts at technology evaluation with the FDA's more rigorous standards. When the FDA rejected this attempt, the plastic surgeons perceived the rejection as based on misunderstanding and prejudice, the specters that had haunted the profession from its inception. Similarly, the ASPRS viewed groups like CTN not as a social-political movement but as conglomerations of dissatisfied patients. In attributing their complaints to improper patient selection, the plastic surgeons attempted to refute the validity of their "horror stories" as evidence.

7

A CALCULUS OF RISK

The FDA was a central player in the public drama of the breast implant controversy, serving as a catalyst to much of the action. In substance, the controversy revolved around the complexities of what I call the agency's calculus of risk. These complexities included struggles to define the nature of valid evidence, the FDA's emphasis on caution, and the delicate balance between the agency and other societal institutions. Beneath the issues ostensibly in question during the controversy—arguments over the safety and effectiveness of silicone breast implant technology—were moral and social disputes. In a broad interpretation, the entire controversy may be read as a reaction by the FDA to the political forces that were threatening its very existence. This chapter places the FDA in its historical context and traces the agency's construction(s) of the problem of implants.

Passage of the first federal food and drug law, in 1906, is widely credited to two men—one a chemist with the U.S. Department of Agriculture, and the other a muckraking journalist. The first, Harvey Wiley, was chief of the department's Division of Chemistry in the late nineteenth century. He undertook a series of reports about contaminants in the food supply and lobbied hard for the passage of federal legislation to correct the problems he found.[1] The other was Upton Sinclair, a socialist whose agitprop novel, *The Jungle,* an exposé of the Chicago meatpacking industry, was published in 1906. While Sinclair's intention had been to reveal the exploitation of workers by the industry, what drew the public's attention, and outrage, was his description of the unsanitary conditions in which meat was processed.

The Food and Drug Act of 1906 regulated the manufacture and interstate distribution of food and drugs. Regulation focused on adulteration and misbranding. The rules against adulteration encompassed unintentional contamination as well as the deliberate substitution of one ingredient for another. The provisions against misbranding aimed to ensure that food and drug labels were accurate representations of their package contents. A finding of adulteration or misbranding was grounds for seizure of the product.[2] The constitutional basis of the 1906 act was the federal government's authority to regulate interstate commerce.[3] Administration of the act was shared by the Departments of the Treasury, Agriculture, Commerce, and Labor. The Agriculture Department's Bureau of Chemistry (the new name for Wiley's Division of Chemistry) was responsible for compliance testing.[4]

The next twenty-five years saw a series of incremental attempts to strengthen the law. For example, in 1912, the Sherley Amendment addressed the problem of false claims about drugs, prohibiting "false and fraudulent curative or therapeutic claims on a label."[5] (In practice, however, because a finding of fraudulence required proof of intent to deceive, the amendment proved difficult to enforce.) Other changes were administrative. In 1927 the Bureau of Chemistry was renamed the Food, Drug, and Insecticide Administration. In 1931, the name was changed again, to the Food and Drug Administration. For the time being, it continued under the aegis of the Department of Agriculture.

Broad revision of the food and drug law did not take place until 1938. Once again, the passage of legislation was preceded by a public outcry over a highly publicized incident. This time it was the "Elixir of Sulfanilamide" disaster of 1937. The Elixir, a widely advertised patent medicine, had been approved by the FDA as unadulterated and properly branded. Unfortunately, one of its main ingredients was a poison—diethylene glycol—and several users died after taking the tonic. Those calling for reform of the food and drug law used the tragedy to point out how the FDA lacked the authority to provide real protection for the public.

The 1938 Food, Drug, and Cosmetic Act greatly increased that authority. The act prohibited the distribution of any drug that had not been tested for safety. This provision instituted a new procedure in which industry was required to submit scientific reports of safety tests to the FDA for premarket approval. Unapproved new drugs could be used only under exemptions for investigational use granted to individual practitioners. The act allowed existing drugs to remain on the market, but if concerns about a drug's safety arose, the FDA was authorized to initiate court proceedings to prove the drug unsafe and have it removed from the market.

In the years that followed, through passage of a number of amendments, the authority of the FDA was extended and clarified, and its administration was moved from the Department of Agriculture to the Federal Security Agency. (In 1953 that agency was renamed the Department of Health, Education, and Welfare, and in 1979 it became the Department of Health and Human Services.) In 1962, with passage of a series of drug amendments, the FDA was authorized to require manufacturers to provide proof of effectiveness, as well as safety, in their premarket applications. "Approval was now conditioned upon the showing of 'substantial evidence' of efficacy, and the burden for proof rested with the manufacturer."[6] In addition, the legislation granted the agency the power to develop guidelines for safe manufacturing practices, to inspect the premises of drug manufacturers, and to suspend the sale of any product believed to be harmful. In response to the Thalidomide tragedy in Britain—a tragedy largely averted in the United States because of the actions of the FDA—the 1962 legislation also included new FDA controls over the distribution and use of investigational drugs.

The new legislation created an agency with both scientific and legal functions. The scientific functions included requirements that the FDA review scientific data, perform some scientific testing, and justify its regulatory decisions based on scientific evidence. The law enforcement functions authorized the FDA to act through the Department of Justice to seize adulterated or misbranded products, to prosecute manufacturers for violating the food and drug laws, and to seek injunctions to prevent violations. These enforcement responsibilities required the agency to deal with legal evidence and courtroom standards of proof.

Certain gaps in the agency's authority remained, however. One such gap was in the area of medical devices. The FDA's power to regulate medical devices, granted by the 1938 act, was limited to ensuring that their labels were accurate, that they were not somehow adulterated, and that they were manufactured under sanitary conditions. In the years before the Second World War, there were relatively few medical devices in use. After the war, however, there was an explosion in the development of new devices. Although many of these devices were fraudulent or even dangerous, the FDA had no authority to stop their sale or distribution. As legislation increased the agency's authority over drugs, it attempted to stretch that power to cover some devices. For example, the FDA managed to gain some control over silicone injections by defining the material as a drug.[7]

Such machinations were cumbersome, however, and thus too uncertain in their result. In 1969, President Nixon, responding to congressional concern, formed the Cooper Committee, a group of experts charged with reviewing the status of medical device regulation. The committee's report, released in 1970, recommended new legislation that would give the FDA an authority over devices analogous to that which they possessed over drugs. The report was highly specific in its delineation of the recommended contents of legislation, including, for example, the provision for three regulatory classes. The committee suggested that, in anticipation of passage of such legislation, the FDA complete an inventory of medical devices currently on the market. It also suggested that the agency form expert advisory panels to review the safety and effectiveness of these devices so that they could be quickly classified after passage of the law. The FDA followed these recommendations. The inventory, released in 1971, found more than one thousand manufacturers producing around eight thousand different devices. FDA advisory panels, constituted around medical specialties, were recruited to review these devices and make classification recommendations.[8]

The Cooper Committee's recommended legislation then languished for several years. True to pattern, it was taken up by Congress only after another highly publicized incident provoked public outrage: passage of the Medical Device Amendments of 1976 followed the Dalkon Shield fiasco, in which the inadequately tested contraceptive device was found to cause serious illness and death.[9]

By the late 1970s, when the agency took up the breast implant issue, the FDA was a huge organization. It was responsible for ensuring the safety of all foods (except for meat, which had remained under the aegis of the Department of Agriculture) and cosmetics, and the safety and effectiveness of all drugs and medical devices sold in the United States. This responsibility encompassed both review and approval of new products, and the monitoring and continued regulation of products already on the market.

Responsibility for medical devices lay with the FDA's Center for Devices and Radiologic Health (CDRH), one of three centers within the agency. Like the other centers, the CDRH used advisory panels of experts, panels usually constituted along disciplinary lines, to review manufacturers' applications and to make recommendations about the disposition of new (and existing) products. By the early 1990s, the CDRH had eighteen such advisory panels; among them was the General and Plastic Surgery Devices panel, charged with review of the breast implant issue.

From its beginnings, the FDA has existed in a careful balance with five other institutions: Congress, industry, physicians, consumers, and the media. Criticism of the agency has been constant and often has focused on an imbalance in one or another of these relationships.

Congress drives the FDA in two ways: it passes the laws that define the agency's authority and responsibility, and it allocates the funds from which the agency derives its operating budget. In addition, several congressional committees are charged with formal oversight of the FDA. (In the House of Representatives, that responsibility rests with the Human Resources and Intergovernmental Relations Subcommittee, the subcommittee that during the controversy was chaired by Weiss.) Because it is a political body, both the legislative and allocative functions of Congress are easily swayed by outside pressures. As the legislative history of the FDA reveals, Congress has often acted to increase the agency's authority in reaction to very public debacles with food and drugs. Increased responsibility has not always resulted in increased funding, however. For example, the agency's long delay in classifying breast implants has been blamed on an overworked staff spread too thin by inadequately funded demands.

Pressure on Congress also decreased the FDA's authority. During the Reagan-Bush era, especially the early 1990s, joint efforts by industry and an executive branch infused with a spirit of deregulation prompted a movement to reform the food and drug laws so as to restrict the power of the agency. As noted earlier, the politics of deregulation was part of the context in which Weiss held his hearing on breast implants.

The relationship of the FDA to the industries it regulates is even more complex. On one level the relationship has been marked by constant struggle over the innovation- and profit-dampening effects of increased regulation. Such criticism was tendered by industry at each point in history when the FDA was granted more authority, and was a focus of Vice President Dan Quayle's Council on Competitiveness, a group of high-level government officials who met during the early 1990s and served as advisers to the deregulation efforts of the Bush administration. Critical analysts have suggested that this struggle may be the inevitable result of the way the agency functions. As one author noted, the FDA "enjoys a power over the industries it regulates that may be clearly defined by statute, but is unenforceable by statute alone. . . . It requires the implicit cooperation of the industries under regulation, and the history of the FDA may be seen as an ongoing attempt on the part of the agency to secure that cooperation, and an ongoing reluctance on the part of industry to grant it."[10]

Other evidence, however, points to a more symbiotic relationship. As an industry-sponsored paean to the FDA, on the occasion of the agency's fiftieth anniversary, declared: "What is good for the consumer is good for business."[11] That is, the FDA's efforts to ensure the public safety has increased public confidence in consumer goods and thus increased the market for those goods. A pharmaceutical manufacturer argued that the existence of the FDA has actually promoted innovation on the part of the drug industry because the agency "has necessitated and even encouraged collection of vastly increased data about medicinal products."[12] And an attorney noted that FDA regulations may level the playing field for fair competition between manufacturers by setting minimum standards that serve to drive out "destructive product competition."[13] The FDA, in turn, is dependent on industry. For example, in the regulation of medical devices the agency must rely upon industry to develop the performance standards that form the heart of Class II regulation.

This close relationship has been the subject of much criticism and was of particular concern in the critiques of the agency that emerged in the 1970s.[14] At a 1974 conference on the safety of food additives, for example, one participant asserted that the "pressures of the system" had turned the FDA into "the agent of the producer."[15] Another argued that "the system of checks and balances . . . is absent. Citizens and consumers . . . are virtually excluded. . . . the concept of matching benefits against risks has been applied to maximize short-term benefits to industry even though this may entail minimal benefits, maximal risks, and externalized costs to the consumer."[16]

Another observer suggests that these seemingly contradictory criticisms of the FDA indicate a pattern of historical evolution in the relationship between the regulator and the regulated: "The attitude of a regulatory agency toward the industry under its jurisdiction undergoes a metamorphosis, changing gradually from initial hostility to a spirit of accommodation and finally to protective concern with the industry's well being."[17]

The FDA's direct relationships are with Congress and industry, but the agency also is influenced by indirect relationships with the users of the products it regulates: physicians and lay consumers. The relationship between the FDA and the medical community has long been one of ambivalence. The FDA serves the interests of physicians in several ways. The agency is part of the legal and political structure that ensures the monopoly control of licensed physicians over prescription drugs.[18] The package inserts mandated by the FDA are an important source of information and education for the physician.[19] And, perhaps most important, the FDA provides a measure of protection against liability: "The physician can use a new drug with confidence and with relative safety, rather than assuming the risk of early experimental use."[20] As the heavy representation of plastic surgeons on the early breast implant advisory panels demonstrates, the FDA recognizes, and relies upon, the expertise of medical professionals.

Although the FDA has no authority to regulate the physician's practice, its jurisdiction over drug and device manufacturers at times has the effect of constraining the physician's treatment options. It is around this issue that the medical establishment and the agency have clashed. The AMA supported drug legislation until the late 1950s, when the agency began to move toward the 1962 amendment that would require manufacturers to prove efficacy. The AMA objected to the efficacy requirement, arguing that "any judgment about this factor can only be made by the individual physician who is using the drug to treat an individual patient,"[21] and fought for repeal, calling it an "incursion of government into the therapeutic relationship."[22] While the FDA may restrict the sale of certain products, any drug or device that has been approved for one use and is available on the market may be prescribed by physicians for any unapproved use as well ("off-label" use). In addition, the FDA's authority extends only to the commercial manufacture and distribution of products. At the height of the implant controversy, one plastic surgeon suggested that the agency had no authority to prevent individual surgeons from producing their own implants for use in their own patients.[23]

The American public has long held a split image of the FDA. While the agency has been portrayed as an organization of Arrowsmiths—dedicated to science, truth, justice, and the American Way[24]—it has also been the target of public criticism for its zeal, or, alternatively, for its laxness.

The criticism of excessive zeal has centered on the argument that the FDA is paternalistic. Not only do the agency's strict requirements act as a disincentive to innovation, but they restrict the rights of citizens to make choices about their own care. For example, Laetrile—touted as a cure for cancer—was never approved for use in the United States. American citizens wishing to use the drug were forced to leave the country to seek treatment. At the height of its popularity, in the late 1970s and early 1980s, Laetrile activists used the paternalism argument to mount campaigns that resulted in passage of state laws decriminalizing the drug.[25]

The criticism of laxness, on the other hand, emerges when the FDA fails to protect the public. Such cases (e.g., scares over food contamination) tend to be attributed—by the agency—to inadequate funding for enforcement or—by outside critics—to the close ties between the agency and the industries it regulates.

In the 1970s, the public criticism of the FDA began to focus on the secrecy with which the agency functioned, in particular, on the lack of consumer representation on its decision-making bodies. FDA addressed the concerns of consumer activists by reconstituting its advisory panels to include a consumer representative, opening many hearings to the public, and making more transparent the agency's deliberative processes. In the 1980s and 1990s, the direct power of consumers increased. Demonstrations and political actions by AIDS activist groups like ACT-UP are, for example, credited with forcing the FDA to streamline drug approval procedures for certain fatal diseases.[26]

The FDA's relationships with Congress, industry, physicians, and consumers are highly interrelated and interdependent.[27] The FDA, charged with the oversight of industry, is itself subject to oversight by a variety of institutions: Congress, the media, and the financial community. Congress, in turn, is vulnerable to pressure from both industry, which makes hefty campaign contributions, and constituents who are also the consumers of drugs and medical devices. Physicians resent intrusion into the doctor-patient relationship but also recognize that public confidence in the medical profession is tied to confidence in the FDA. The media provide a forum in which all of these groups struggle to have their claims heard. Through their own initiative, the media also become actors in bringing certain explosive issues (the meatpacking industry, Elixir of Sulfanilamide, the Dalkon Shield) to public scrutiny.

At the heart of the criticism of the FDA is a tension surrounding the calculus of risk. Risk is in question in most conflicts between the FDA and the institutions it is linked to: the balance between potential risk and corporate growth; the amount of risk an individual may choose to assume without regulatory interference; the outcry when risk becomes harm. Interpreting and managing risk are the central functions of the agency. The silicone breast implant issue provides an example of this calculus in action.

Robert Crandall and Lester Lave described the four major issues underlying government regulation of health and safety risks: (1) Is there a scientific foundation for government action? (2) Will setting a standard reduce risks? (3) Is there enough scientific evidence to assess the risks and benefits of standards? and (4) Has sufficient attention been paid to uncertainties?[28] The calculus that the FDA uses to interpret and manage risk engages each of these questions. In setting standards and making policy decisions, the agency must assess the evidence of safety and effectiveness, weigh risk and benefit, and judge the sufficiency of information.

The Medical Device Amendments of 1976 authorized the FDA to regulate medical devices for safety and effectiveness. The legislation, however, provided little explicit definition of these concepts, noting only that safety and effectiveness "are to be determined A) with respect to the persons for whose use the device is represented or intended, B) with respect to the conditions of use prescribed, recommended, or suggested in the labeling of the device, and C) weighing any probable benefit to health from the use of the device against any probable risk of injury or illness from such use."[29]

Several years later, in an explication of the procedures used to classify devices, the FDA published an interpretation of the law that set out its working definitions of safety and effectiveness. Safety is "valid scientific evidence that the probable benefits to health from use of the device for its intended uses and conditions of use, when accompanied by adequate directions and warnings against unsafe use, outweigh any possible risks."[30] Effectiveness exists when "in a significant portion of the target population, the use of the device for its intended uses and conditions of use, when accompanied by adequate directions for use and warnings against safe use, will provide clinically significant results."[31]

An observer of the FDA described three stages in the assessment of safety: "objective, scientific determination of the discernible effects" of a product; "judgment about which of these effects is a risk and which is a benefit"; and "the decision that the agreed upon benefits . . . exceed . . . [the] agreed upon risks."[32] The FDA uses several mechanisms to work through these stages: technical review of premarket applications; enforcement of manufacturing and testing standards; postmarket monitoring for adverse effects; analysis of failed products; basic scientific research; and advisory panels.[33] While these mechanisms are easily understood as attempts at "objective, scientific determination," it is more difficult to see how they allow the agency to weigh risk and benefit.

The language used in the agency's interpretive documents sheds some light on that weighing process. Parsing the definitions of safety and effectiveness quoted above reveals two important emphases: the quantitative nature of the assessment (conveyed through words like "probable" and "significant") and the reliance on "valid scientific evidence." Further explication of the "valid scientific evidence" describes it as "evidence from which it can fairly and responsibly be concluded by qualified experts that there is a reasonable assurance of the safety and effectiveness of a device. . . . The evidence required may vary according to the characteristics of the device. . . . Valid scientific evidence includes well-controlled investigations, partially controlled and uncontrolled studies, well-documented case histories by qualified experts, and, in some cases, reports of significant human experience of a marketed device."[34] The determination of whether particular evidence is valid scientific evidence rests with the FDA commissioner.

While the FDA requires valid scientific evidence to prove safety, the standards for suggesting potential risk are quite different: "Isolated case reports, random experience, reports lacking sufficient details to permit scientific evaluation, and unsubstantiated opinions are not regarded as valid scientific evidence to show safety or effectiveness. Such information may be considered, however, in identifying a device the safety and effectiveness of which is questionable."[35]

Such passages highlight several important aspects of the FDA's calculus. First, in its hierarchy of assessment, the agency privileges both "qualified experts" and certain standards of scientific validity (e.g., controlled studies). These emphases imply that the weighing of risk and benefit is a scientific function, the province of experts and not laypeople. Second, the language describes an affirmative duty to prove safety. That is, the FDA requires positive demonstration of safety, not just a lack of evidence of harm. As the agency has argued: "Congress intentionally placed upon industry the burden of furnishing sufficient evidence to substantiate the safety and effectiveness of a device. . . . the absence of such data may be the basis for classification of the device into Class III."[36] Risk, on the other hand, need never be proved. It is enough that it simply be suggested.

Observers of the FDA have suggested that this emphasis on affirmative proof of safety is the defining feature of the agency. The agency's mandate, they argue, is to avoid type II errors—that is, to avoid approving a product that later is found to be unsafe. This orientation places the FDA at odds with the industries it regulates, which argue that in avoiding type II errors the agency commits too many type I errors (rejection of safe products).[37]

But the language also suggests the extent to which the weighing of risk and benefit is a "trans-scientific" function. That is, it "is a public, not a scientific, decision."[38] Clues to the trans-scientific nature of risk-versus-benefit assessment lie in the use of terms like "reasonable assurance" and "sufficient evidence." "Reasonable" is both a scientific standard—linked to the concepts of validity and statistical significance—and a legal standard. The FDA is an institution that operates in both realms. Thus, while the term may refer to scientific decision making, there is an implication that this "reasonable assurance" must also pass muster in a legal setting. "Sufficient evidence" points to the same interlinking of the scientific and the legal.

The rhetoric of sufficiency introduces the idea of uncertainty. Uncertainty exists when there are inadequate data, when there is a lack of scientific agreement over the interpretation of data, and when there is no public consensus over risk.[39] The language of "insufficient evidence" may refer to any or all of these situations. Uncertainty is a necessary condition for controversy to exist.[40] The FDA's emphasis on the sufficiency of evidence reflects two interpretations of the word "sufficient." One refers to quantity. When there is an insufficient quantity of evidence, there is no basis upon which to make a determination of safety and effectiveness. "Sufficient" also refers to quality or type of evidence. Evidence may also be insufficient if it does not meet the standards of "valid scientific evidence." The FDA has argued that "in the absence of safety and effectiveness data, it may be impossible to determine that no . . . potential risk exists."[41] Thus, given the affirmative duty to prove safety and the agency's emphasis on the avoidance of type II errors, an insufficiency of information must lead the FDA not to determine that there is a risk, but to function in a state of uncertainty in which its actions may be identical to those taken in the case of assumed risk.

Another source of uncertainty lies on the trans-scientific nature of risk assessment. The FDA, as an institution of scientism,[42] emphasizes the scientific—objective, quantitative—nature of its calculus of risk. Science alone, however, can never be determinative of regulation. Although scientists may "help in the formulation of the right questions, the search for the best data and judgments, and the identification of the range of uncertainty,"[43] the ultimate choices about how much and what type of risk should be assumed are social. Conflicts over risk that seem to be arguments about technical questions are, at bottom, really clashes over values.[44] Uncertainty, then, derives not only from differences over the interpretation of scientific evidence (or even from disagreements over what constitutes valid evidence) but from the multiplicity of moral standpoints that exist among the institutions of modern society.

Following passage of the Medical Device Amendments of 1976, the FDA began the long process of classifying the thousands of devices that were already on the market. Breast implants were one of those devices. The FDA brought the issue of implants to its General and Plastic Surgery Devices panel as early as 1978, but it was not until 1982 that the panel issued its recommendation. As described in chapter 6, the panel, with its heavy representation of plastic surgeons, recommended that implants be placed in Class II (the regulatory classification that would not have required the manufacturers to provide any proof of safety or effectiveness.) The FDA, however, disagreed with its panel's recommendation, announcing that it would instead place implants in Class III.[45] The agency supported this decision with two claims: first, that implants represented a potential risk to health; and second, that there was insufficient information about the safety and effectiveness of the devices to allow them to be placed in a less restrictive regulatory category.

To support its claim of potential risk, the FDA listed a number of specific complications, including leakage of silicone gel, migration of silicone in the body, infection, capsular contracture, and interference with tumor detection. A review of the literature had shown, the agency wrote, that a "significant portion" of breast implant recipients experienced at least one of these adverse effects. In addition, the agency expressed concern about the possibility of implant rupture following closed capsulotomy and about the lack of knowledge of the exact chemical composition of the silicone gel and its possible toxicity. In sum, the FDA had determined that there was "insufficient information" about these risks for general controls or performance standards (i.e., Class II restrictions) to "provide a reasonable assurance of the safety and effectiveness of the device."[46] Because the device "present[ed] a potential unreasonable risk of injury," it should be placed in Class III.[47] The FDA supported its classification decision by noting that there were "ongoing scientific debates" about the safety of implants.

Here, the FDA's use of the term "sufficiency" referred both to type and amount of information. With this notice of intent, the agency argued that while the quantity of its information was sufficient to suggest risk, it did not have enough information in the form of valid scientific evidence to allow a less restrictive classification. The evidence proving safety was lacking as to both amount and type. The agency thus constructed the problem of implants not as one of harm—although it made claims about the possibility of danger—but as one of uncertainty. The risk posed by implants was "unreasonable" not because it was high but because it was unknown. The fact that there were "scientific debates" about the possible risk bolstered this construction of the problem.[48]

Reaction to the FDA's notice of intent to place implants in Class III challenged both of the agency's primary claims. Implant proponents used history, experience, and scientific counterclaims to argue that there was no evidence for harm caused by silicone. (The plastic surgeons, for example, argued against each of the specific risks cited by the agency: e.g., capsular contracture was "not a complication, but rather a normal event in *all* wound healing";[49] data on tumor detection suggested that implants made early detection "*even easier*"; there was "no support" for the claim that silicone gel might be toxic.) These implant proponents accused the agency of setting an "impossible standard" by demanding that a manufacturer "show an absence of risk." Such a standard was particularly odious given the fact that the "FDA will, in evaluating safety, consider such data as '. . . isolated case reports, random experience, reports lacking sufficient details to permit scientific evaluation, and unsubstantiated opinions.'"[50] Because it had not considered the benefits of breast implants, the FDA had failed to complete a formal risk-versus-benefit analysis. This "failure to examine critically the evidence on which it relies and to describe that examination in detail . . . constitutes arbitrary and capricious action." Thus the agency's actions were "legally deficient."[51]

The FDA responded to its critics in the 1988 notice of its final classification of implants. Here, the agency answered specific counterclaims (e.g., citing "recent scientific data" which "reveal occurrence of allergic reactions, silicone lymphadenoma, [and] morbidity due to silicone . . . with unknown long-term effects")[52] and laid out a clear exposition of its own conceptualization of risk-versus-benefit analysis, arguing that its actions had in fact been guided by law, not caprice.

The explanation of the FDA's procedures for weighing risk and benefit elucidated some of the issues called into question by the opposition, in particular, the apparent difference in standards for proving safety and suggesting risk. First, the agency noted the statutory requirement that all implantable devices be placed in Class III unless there was affirmative evidence of safety.[53] This requirement for affirmative evidence placed the burden of proof squarely on the manufacturers, it argued. Second, the proponents of implants had so far failed to prove the benefits of the devices scientifically. "Any risk of illness or injury is unreasonable when no evidence is available of probable benefit to the health of those persons for whose use the device is intended," the agency explained. [54] A finding of "'potential unreasonable risk of illness or injury' ha[d] two significant features": "First, the requirement that a risk be unreasonable contemplates a balancing of the possibility that illness or injury will occur against the benefits from use. Second, the risk need only be a potential one. The risk may be one demonstrated by reported injuries or it may simply be foreseeable. The fact that a device is being marketed without sufficient testing is an adequate basis for the . . . conclusion that the device presents a potential unreasonable risk to health."[55] A potential risk would be unreasonable—and thus cause enough for Class III placement—as long as there was uncertainty surrounding its potentiality and as long as benefit, too, continued to be unproven.

In response to the FDA's stated intent to require premarket applications for breast implants, in January of 1989 Dow Corning submitted some data on implants, including internal reports of customer complaints and product returns, to the FDA for a "presubmission review." The data were given to the agency under a mutual understanding that they would not become a part of the manufacturer's official application unless they could be held confidential. The data that were not exempt from revelation (i.e., those that the agency would not agree to keep secret) were to be returned to Dow Corning.[56] Later that winter, Public Citizen learned of the existence of these documents and asked that the FDA release them to the public. When the FDA denied its request, Public Citizen sued the agency.[57]

On November 20, 1990, a federal judge ordered the FDA to release the documents. The FDA had argued that because the manufacturer's submission had preceded the official requirement for premarket applications, the agency was under no statutory requirement to make the contents of a "voluntary" submission public. In fact, the FDA asserted, such a revelation would make other manufacturers reluctant to submit such information to the agency for review.[58] The judge disputed this reasoning, noting that to accept it would be to make the FDA "the most *impotent* agency in Washington . . . [by implying that] the only way it can protect the public's health and welfare is by begging manufacturers of possibly lethal devices to submit information voluntarily. While the FDA may presently rely on so-called voluntary submissions, this is clearly due to the agency's ineptness and to its failure to pursue avenues that are open to it."[59] Soon after this decision, the commissioner of the FDA, Frank Young, resigned from the agency. Several months later, David Kessler, trained both as a physician and a lawyer and a former staffer for Utah Republican senator Orrin Hatch, was nominated to the position by President Bush and confirmed by the Senate.

The episode highlighted some of the vulnerabilities in the FDA's political position. Any intervention by the courts served to make the agency look weak. The clear implication of the judge's decision was that the FDA wielded little actual power—that, as consumer advocates like Public Citizen had argued for years, the agency was often little more than a rubber stamp for industry. Such evidence of weakness provided ammunition for opponents of the FDA from both ends of the political spectrum: those who argued that the agency was unnecessary and those who argued that the agency was ineffective after years of stewardship by deregulationist conservatives.

The FDA's bureaucratic process ground on. In the 1990 proposed rule to require premarket applications for silicone breast implants, the FDA again presented its claims for what was known about the risks and benefits of the devices. The specific risks it cited included capsular contracture, silicone gel leakage and migration, interference with early tumor detection, carcinogenicity, teratogenicity (a possible cause of birth defects), and—for the first time—immunological sensitization and autoimmune disease. Explicitly, the agency indicated that the manufacturers' applications would have to address these risks. On the issue of benefit, the FDA conceded the definition of benefit as psychological, citing some of the psychological studies of augmentation and reconstruction discussed in chapter 4, but it called for more data on "patient satisfaction, improved self-image, and improved outlook."[60] The demand for more data on both physical risks and psychological benefits acknowledged the dual meaning of "sufficient information" and the different standards for proving safety and suggesting risk: "There is reasonable knowledge for the risks and benefits associated with the silicone gel-filled breast prosthesis. There is [sic], however, insufficient valid scientific data to permit FDA to perform a risk/benefit analysis."[61]

The FDA's statements indicate that despite shifting concerns about specific risks—concerns that were the subject of internal disputes among FDA staff scientists—the agency continued to frame the problem of implants as one of uncertainty. Its statements about sufficiency of information and valid scientific data were meant to be read as imperatives for the manufacturers, but they also, implicitly, indicated that the agency was claiming its own authority as the arbiter of final decisions.

Public comment received by the FDA in response to the 1990 notice—the agency reported receiving some 2,670 communications—made three kinds of claims: claims about silicone, claims about the benefits of the devices, and claims about the FDA's handling of the implant issue. Predictably, the claims about silicone emphasized the safety of the material. The claims about benefit charged the agency with underestimating the value of the devices to women who received them. (In what was probably part of a coordinated campaign, the FDA received more than 2,500 letters describing "the psychological and psychobiological benefits of breast prostheses . . . [and] stat[ing] that it is important that women be given the chance to freely choose silicone gel–filled implants as an option as long as they are well informed of the benefits and risks of the surgery.")[62]

The claims about the FDA questioned the competence and motive of the agency. "FDA has utilized old, unrelated anecdotal evidence or unsupported opinion. . . . FDA has misread, misquoted, acted in a biased and unreasoning manner, and utilized information not appearing in the administrative record, thus acting arbitrarily and capriciously," one critic wrote.[63] Others questioned the addition of new risks to the FDA's list of potential risks and asserted that the manufacturers were not being given enough time to comply with the FDA's requirements for the premarket applications. One comment used historical arguments to question the agency's methods: "Congress intended that less rigorous scientific evidence should be applied to 'old' devices. . . . FDA should accept meaningful scientific data from the 20 years of use of the device rather than require prospective clinical data."[64]

Substitution of the word "meaningful" for the word "valid" touched the heart of the dispute: Whose expertise was to be privileged? Was it the clinical experience of plastic surgeons? The personal experience of satisfied recipients? The suffering of silicone victims? In insisting on "validity" as the standard of evidence, the FDA was asserting its own authority over the implant issue by claiming the disinterested objectivity of scientism and constructing the problem as one of uncertainty. The idea of "meaningful" evidence, however, which opened the assessment process to other groups and epistemologies, called into question not only the degree of uncertainty but the FDA's claim of disinterested objectivity. The problem of implants, then, was reconstructed by implant proponents as one of the FDA's arrogance: the agency simply misunderstood implants. By its actions it was threatening individual freedoms. In view of such a threat, the burden of proof should shift back to the FDA. If implants were to be so strictly regulated, it should first be up to the agency to prove harm.

This shift to challenging the FDA's authority (rather than simply challenging its interpretation of the evidence) was a reflection of the broader criticism being directed at the agency. During the period from 1991 to 1992, the very height of the implant controversy, the FDA was at the center of another controversy, this one involving Vice President Dan Quayle's Council on Competitiveness and its efforts to promote deregulation at the federal agencies charged with safeguarding public health and safety.

Formed in 1990 as the successor to President Reagan's Task Force on Regulatory Relief (which had been headed by then Vice President George Bush), the Council on Competitiveness was composed of six permanent members: John Sununu, then Bush's chief of staff; Richard Darman, then director of the Office of Management and Budget; the secretaries of the Commerce and Treasury Departments; the chairman of the Council of Economic Advisors; and the attorney general. The council staff was drawn from the business world and included a number of well-known conservative activists.[65] The president's charge to the council was to continue the deregulatory efforts that had been one of the hallmarks of the Reagan administration (efforts his own administration had been accused of relaxing).[66] Most of the council's work focused on revising agency-level rules (the standards and practices federal agencies write to operationalize the legislation passed by Congress) believed to be too burdensome to industry. Democratic politicians and consumer activists objected to both the council's process and to its results, charging that it was a kind of "shadow government," a puppet of big business, and alleging conflict-of-interest charges against individual members.[67]

In November of 1991, the FDA, under the leadership of Commissioner David Kessler, announced a sweeping proposal to reform the way in which the agency reviewed and approved new drugs. Billed as a joint effort of the agency and the Council on Competitiveness, the proposal called for opening up the review process to private contractors, revising the way in which effectiveness was defined and measured, changing liability law to make it harder to sue drug companies, and instituting a kind of approval reciprocity with a number of foreign countries (i.e., drugs that were already approved elsewhere could be sped through the process in the United States).[68] Touted as a way to reduce the time and costs to industry needed to gain approval for new drugs by making more efficient use of FDA staff and resources, the proposal was a response both to the deregulatory philosophy and the demands of health activists, like ACT-UP, for a streamlining of the drug approval process.

Almost immediately, however, the proposal was attacked as a serious weakening of the FDA's ability to protect the public. Public Citizen's Sidney Wolfe argued that it "could be a major step backwards."[69] Congressman Weiss, chairman of the House subcommittee charged with oversight of the agency, expressed his suspicion that, rather than being a collaboration, the council had in fact strong-armed the FDA into endorsing its own deregulatory agenda. Weiss immediately launched an investigation into the process through which the reform measures had been designed. (Eventually, documents produced by the FDA revealed that there had been a good deal of internal dissension about the reforms.)[70] In the spring of 1992, the FDA implemented a somewhat watered-down version of the reform proposal.[71]

This ongoing clash between Democrats and consumer advocates, on the one hand, and conservative, business-minded deregulationists, on the other, formed the backdrop to the FDA's role in the public drama of implants. During his 1990 hearing on breast implants, for example, Weiss presented the problem of implants as one of inadequate scrutiny, a criminal oversight by an agency weakened by Republican deregulatory cost cutting. To make his argument, of course, Weiss had to assume the danger of implants. Similarly, implant proponents framed the problem of implants as one of too much government regulation, an intrusion into the lives of free Americans. Such an argument required an assumption of safety. Caught in the middle, the FDA had to defend itself against both the anti-deregulationists—whose political position rested on an assertion of the agency's impotence—and the deregulationists—who would strip the agency of even more power.

In the 1991 notice that set a deadline for the receipt of premarket applications from the implant manufacturers,[72] the FDA first addressed some of the technical questions raised by its critics. For example, the agency noted that the manufacturers had had ample time to prepare their premarket applications; after all, it had been nearly a decade since the FDA had announced its intention to put implants in Class III. Quickly, however, the text's focus shifted to making claims about the agency's challenged authority. The notice emphasized the legal basis of the FDA's authority and described the agency's scrupulous adherence to procedure. It rejected the attempt by implant proponents to shift the burden of proof to the agency by repeating the distinction between proving safety and suggesting risk. The agency's rhetoric centered on its refrain of "sufficiency" of information, pointing to its continued construction of the problem of implants as one of uncertainty.

An FDA Talk Paper about the notice also emphasized the problem of uncertainty, arguing that it was the manufacturers' responsibility to provide the agency with the scientific data necessary to remedy it.[73] The document drew a distinction between "known problems"—"infection, false mammography results, silicone leakage, and implant failure"—and "potential unknown long term risks such as immune reactions and carcinogenicity." The manufacturers' application would have to provide "information . . . on the severity and the incidence of these existing or potential adverse effects." The manufacturers now had ninety days to submit their applications.

As described in chapter 1, the agency's review of the breast implant manufacturers' applications was informed by scientific judgment but also by political concerns. The FDA apparently recognized that outright rejection of the applications—which its internal review had judged largely inadequate—would result in a protracted battle. In the course of such a legal wrangle, the opportunity to collect more data would be lost and the FDA would run the risk of seeing its authority further degraded by a loss in court. Thus, despite "major methodological flaws," the agency agreed to bring the applications of four manufacturers to its advisory panel.

The public setting for the application review was the three-day hearing held in November of 1991. The FDA began the meeting by asserting its own claims. First, the agency made it clear who bore the burden of proof: "The FDA is not required to prove that breast implants are unsafe. On the contrary, manufacturers have the responsibility of showing . . . that their breast implants are safe and effective. . . . the burden of proving safety rests with the manufacturer." Second, it outlined the basis upon which safety was to be determined. It explicitly rejected the implant proponents' arguments for relying on the history of clinical use and for viewing the problem as one of individual rights: "Simply saying that devices have been used for years without major problems does not meet the safety standard. . . . The law . . . does not permit FDA to simply leave the decision up to the individual patient. Informed choice is a vital part of any surgical procedure. [But] this alone is not sufficient to fulfill the requirement of the medical device law." What the law did require was an assessment based upon the weighing of known risks and benefits: "An application may be approvable if these risks are well understood and if they are outweighed by the benefits of the device. On the other hand, if the risks are unknown or poorly quantified, the benefits side to the equation must be correspondingly higher in order to justify approval."[74]

In introducing the manufacturers' applications to the advisory panel for its consideration, the FDA made specific claims for the inadequacies of the applications. Agency staffers argued that the applications lacked "certain critical data": information about the physical and mechanical characteristics of implants, the chemical composition of the gel, the metabolic fate of silicone in the body including the potential toxicity and immune effects, the incidence of known adverse effects in the implant population, and the extent to which the devices interfered with mammography. The studies that the manufacturers had submitted to the FDA were hampered by selection bias, short follow-up periods, the lack of an overall denominator for the implant population, and unstandardized measures of benefit. One staffer noted that while the FDA "acknowledges that the materials submitted in these PMAs [premarket approval applications] do not show a hazardous device posing a major threat to the health of users . . . questions remain as to the presence of long-term risk, the accurate measures of the incidences of complications, and the effects of the implants on mammography."[75]

These claims reasserted the FDA's authority over the problem of implants (and, more broadly, over the area of health and safety regulation). They reemphasized the construction of the problem as one of scientific uncertainty, implicitly and explicitly setting out the rules and rhetoric with which the FDA would make a decision.

The manufacturers received the message. Their presentations, which lasted through a day and a half of testimony, reflected a rhetorical strategy of speaking in the scientific language of the FDA. Each manufacturer used panels of scientific experts to testify about the safety of implants and, especially, about the benefits of the devices. Each referred, verbally and—by pointing out the physical piles of documents that they had carried into the hearing room—visually, to the volume of scientific research that they had performed. The manufacturers' scientific claims related to the safety of implants and the benefits of the devices. Other claims, however, attacked the FDA.

The agency had misinterpreted the statutory standard, the manufacturers asserted. "The FDA's charge to this panel is not to require that manufacturers demonstrate 100 percent safety, because that standard does not exist and cannot be achieved by any product. . . . Instead, what the manufacturers must provide is a reasonable assurance of safety and efficacy, associated with the proper use of the product. The information provided in our PMA applications, as well as our testimony here today, provides this reasonable assurance."[76] Breast implants were being held to different, more difficult, standards than other devices, even other silicone devices. For example, Norplant, an implantable contraceptive encased in a silicone shell, had been approved without any questions raised about the silicone.[77] In addition, the agency was being inconsistent in its treatment of those opposing implants and those championing their availability: "Detractors are not to be subjected to the same burden of proof as are the manufacturers. . . . The result is that an inherent bias has developed among many who have been educated by the adversarial position."[78] Finally, the FDA had been unfair in its handling of the applications: "The manufacturers have been subjected to a very inconsistent PMA process."[79]

The manufacturers' claims about the safety of silicone and the benefits of implants conceded the FDA's authority over the devices and attempted to defend them based on the agency's stated criteria for assessing risk and benefit. In general, the manufacturers seemed to accept their responsibility for proving safety, although they attempted to portray that responsibility as fulfilled through the historical demonstration of minimal risk. (An exception to the manufacturers' assumption of responsibility was their argument that closed capsulotomy was the major cause of rupture. In this claim, the manufacturers attempted to shift responsibility to the plastic surgeons. The procedure had been a source of contention in the manufacturer-surgeon relationship since the mid-1980s, when a series of liability suits resulted in a legal determination that the manufacturers had not provided sufficient warnings against the use of the technique. By 1990, while manufacturers officially opposed the procedure as outside "proper use," the plastic surgeons continued to practice it.) The claims about the FDA's ostensible mishandling of the implant issue, however, implied that the problem of implants was one of unfair and unreasonable government regulation. Such an argument, as we have seen, resonated with the contemporary political debate.

The deliberations of the advisory panel members over the applications and the "public health need" for implants revealed how they were trying to integrate several constructions of the problem of implants. Their decisions—to reject the applications on the basis of insufficient information but to allow implants to remain on the market while research continued—reflected their acceptance of the FDA's claims of uncertainty but also similar acceptance of the claims of implants proponents that the history of use should count in making determinations of safety and the highly emotional claims of implant advocates of the "crying need" for the devices.

As explored in previous chapters, the panel's recommendation was quickly followed by the courtroom revelation of the Dow Corning documents. These documents immediately became the focus of the regulators' decisions. Thus, on January 6, 1992, the date by which the FDA was required to make a decision regarding its advisory panel's recommendations, Commissioner Kessler announced that rather than following those recommendations he would instead, on the basis of the newly revealed information, call for a moratorium on the sale and distribution of silicone gel breast implants and ask plastic surgeons for voluntary adherence to the moratorium as regarded their use.

Kessler asserted that despite thirty years of clinical experience, and nearly a decade of FDA involvement, there was still overwhelming uncertainty about the complications and risks of silicone breast implants. "The manufacturers . . . have failed thus far to provide adequate evidence that they are marketing a safe product," he said. He emphasized that the FDA "is not opposed to implants. We recognize the value of these devices. . . . We *want* to see safe breast implants available for all women who need them." However, he argued, "the information we've acquired since the last panel meeting . . . has increased our doubts about these products." The FDA, he said, "owe[s] it to the American people to see to it that these questions are thoroughly investigated."[80]

Kessler's claims portrayed the problem of implants as one of increased doubt, doubt that reinforced the FDA's broader claim of uncertainty. In stressing scientific uncertainty, he reasserted the FDA's ownership of the problem of implants—and its authority to manage a solution. His mention of "the American people" was a reference to the ultimate source of that authority.

In the FDA's opening statements at the February meeting of its advisory panel (called to review the newly revealed information and, if necessary, to make revised recommendations), the agency went into greater detail than it had before in the presentation of evidence. FDA staffers reviewed the Dow Corning documents, voicing the agency's claim that the significance of this evidence lay in its demonstration that "during this period there were significant questions about the quality of the product and manufacturing process, about the rate and consequences of rupture and gel bleed, and about the sufficiency of safety testing of the product and its components. . . . in spite of these questions, thousands of breast implants were manufactured and implanted into women during this period . . . [and] as late as 1983 and 1985 there was a recognition that fundamental safety testing was necessary and had not been adequately performed."[81]

These revelations had prompted the agency to look more closely at issues of quality control as they related to implant rupture and gel bleed. "How often do these devices actually rupture?" asked Kessler. "What happens to these devices 10, 20, or 30, or 50 years after implantation? What is an acceptable device failure rate? Do we know when a silicone breast implant ruptures? How do we screen women effectively and safely for possible rupture? What should be done when an implant ruptures? What are the consequences of rupture, leakage, and bleed?"[82]

The FDA's statements and evidentiary claims continued to revolve around its construction of the problem as one of uncertainty. For example, in emphasizing rupture, Kessler chose an issue that exemplified uncertainty about both safety and effectiveness. Arguably, the agency went further than it had before to explicate its concerns. Despite this effort, however, the agency continued to argue not that implants were dangerous but that there was not sufficient information with which to assess the risk of danger.

The agency's more aggressive presentation garnered equally aggressive criticism. As discussed in chapter 5, when the focus of the hearings shifted to the issue of autoimmune disease, the proponents of implants responded by charging that such claims were the result of "pathologic science." Such "junk science" had its origin not in the laboratory but in the courtroom, where highly paid "experts" used "the science of the unreal and bias" to bolster the groundless legal claims of greedy plaintiffs' attorneys. [83] (Interestingly, tort reform was another focus of the Council on Competitiveness.) Now critics were charging that the FDA had fallen victim to the pathology. Marilyn Lloyd, a congresswoman from Tennessee and a breast cancer patient who had been unable to get silicone implants because of the moratorium, voiced this claim strongly, asserting that the FDA had "demonstrated . . . an inability to deal with this issue on a scientific basis." She explicitly linked the agency's actions to the current political threat, calling the moratorium "more self-serving to an anxious federal agency trying to overcome a rather undistinguished past than it is to protecting the health of American women."[84]

The attack on the FDA's handling of the silicone breast implant issue once again reconstructed the problem of implants: it was one of politics and lack of science. The effect of the larger political debate about the role of the FDA had been to politicize the scientific evidence in this particular case, tainting it in such a way as to render it uninterpretable. This construction challenged both the FDA's legal status and authority and its scientific competence.

On the last day of the advisory panel's meeting, the panel issued its recommendation, which was to allow silicone gel implants to be used for reconstruction in wide access clinical trials and for augmentation in more limited access trials, but to restrict entirely their use outside of trials. The panel's recommendation reflected its acceptance of the FDA's construction of the implant problem as one of uncertainty and its application of the FDA's calculus of risk in placing uncertainty on the risk side of the equation. By following the agency's calculus so closely, the panel also endorsed its legitimacy.

On April 16, 1992, Commissioner Kessler announced the FDA's decision about implants, a decision that was virtually identical to the recommendations made by the advisory panel in February. In making the announcement, Kessler emphasized several points: the agency's recognition of a public health need for reconstruction; the existence of "special problems" for augmentation patients that justified a policy distinction; and the requirement that further study was needed. "Our primary goal," he said, "is to put in place a process to obtain adequate information about the safety of these devices."[85]

The resolution framed the problem as the FDA had from the beginning—as one of insufficiency of information and thus of uncertainty. It maintained the agency's statutory authority. The parameters of the decision—the focus on further study, research, and information gathering—also worked to maintain the agency's scientific authority. (For example, the FDA gave a scientific, rather than a moral, justification for the policy distinction between reconstruction and augmentation.) The solution was formulated, presented, and justified using the rhetoric of science, implicitly a rejection of the charge that the agency had been swayed by political considerations.

But in the months that followed the decision, the FDA continued to be dogged by criticism of its science and challenges to its authority. In June of 1992, for example, the *New England Journal of Medicine* published several editorials about the decision. The first, by *Journal* editor Marcia Angell, criticized the basis of the decision.[86] Pointing to the "disparity between the little that has been demonstrated about the risks and the abundance of anecdotes now being related in the media or the courts,"[87] she argued that the agency had weighed risk and benefit inappropriately, allowing itself to be swayed by anecdotal evidence while giving "short shrift" to the benefits of implants, acting "as though there were none . . . [and] holding breast implants to an impossibly high standard: since there are no benefits, there should be no risk."[88]

The FDA's decision, she continued, was coercive, anxiety producing, and sexist. Women who wanted implants would now be forced into clinical trials, removing the voluntariness necessary for participation in research. The decision had raised women's fears about implant safety without basis, leading to increased litigation as well as personal anguish. In addition, the FDA had ignored "the social context. Targeting a device used only by women raises the specter of sexism—either in having permitted the use of implants in the first place or in withdrawing them. The view that it is sexist to withdraw implants is exacerbated by the fact that people are regularly permitted to take risks that are probably much greater than the likely risk from breast implants; they do so when they smoke cigarettes, for example, or drink alcohol to excess. The argument that many of these other risks lie outside the purview of the FDA may be seen by many as a legalism."[89]

Claiming that there was a "sharp divide" among feminists over the issue of implants, Angell opined that it was "possible to deplore the pressures that women feel to conform to a stereotyped standard of beauty, while at the same time defending their right to make their own decisions."[90] Her solution was to "[permit] women to continue to receive breast implants, regardless of participation in studies and regardless of whether the purpose is augmentation or reconstruction, with the provision that they . . . [receive] the same information as they would in a [clinical] trial."[91]

In seeking to contextualize the breast implant controversy,

Angell was taking an important step. The direction in which she was moving, however, was toward questioning the FDA's claim that its ultimate authority was the endorsement of the American people. Perhaps, she seemed to suggest, the people were on the verge of revoking that endorsement.

An editorial by plastic surgeon Jack Fisher provided another critique. He restated many of the arguments made by the plastic surgeons during the controversy, accused the FDA of insensitivity to the needs of women, and claimed that the agency had fallen under the influence of "the exaggerated claims of consumer-advocacy groups" while "the carefully considered position statements of many respected organizations . . . seem to have been ignored."[92] In contrast to Angell, who had suggested a broadening of FDA accountability to encompass the social context of its decisions, Fisher was emphasizing a more traditional model, one in which the experts—and only the experts—were considered qualified to make scientific judgments about regulation.

The FDA's response to these criticisms came in an editorial by David Kessler that appeared in the same issue.[93] His justification for the agency's decision continued to emphasize the scientific uncertainty surrounding implants, the long "list of unanswered questions" about safety. His explanation of the distinction between reconstruction and augmentation featured an explicit rejection that the decision had been based on "any judgment about values, but rather reflected the FDA's opinion that the risk-benefit ratio does not at this time favor the unrestricted use of silicone breast implants in healthy women."[94] Kessler provided a succinct argument for the authority and legitimacy of the FDA, one which situated the agency in its own "social context": "If members of our society were empowered to make their own decisions about the entire range of products for which the FDA has responsibility . . . then the whole rationale for the agency would cease to exist. . . . The FDA was established as a result of a social mandate. . . . Manufacturers have vested interests. Between these interests and the interests of patients, the FDA must be the arbiter." To make the standard "caveat emptor," Kessler continued, would be "to impose an unrealistic burden on people when they are most vulnerable to manufacturers' assertions."[95]

Several months later, there was more criticism—this time focused on the laxness with which the agency had handled the breast implant issue. Staffers for Congressman Weiss's subcommittee (Weiss himself died in September of 1992) released a report about the history of the FDA's regulation of breast implants.[96] The report implicated both the manufacturers and the plastic surgeons in a deliberate pattern of deception and accused the FDA of complicity. The agency had ignored twelve years' worth of warnings about implants, the report charged. It had ignored the opinions of its own scientists and withheld information from its advisory panels. All of this was due to the influence of pro-industry lobbyists and conservative politicians, who together—and again, with the complicity of the agency—had succeeded in weakening its ability to act in the public's interest.

The FDA's decision about breast implants, announced just a week after the agency had implemented many of the reform measures developed by the Council on Competitiveness, seemed to be a successful assertion of authority in a time of threat to that authority. The agency had triumphed in making its own construction of the problem of implants the dominant one for policy-making purposes. As we have seen, however, the controversy over implants revealed that beyond the political threat of deregulation, the FDA was facing a more general challenge. The breast implant controversy, despite—or perhaps because of—its absurdities, had laid bare the trans-scientific nature of much of the FDA's calculus of risk and the moral and social, as well as political, grounds of its authority.

8

AFTER BABEL
An Epilogue

Attention shifted away from the FDA and the regulatory status of breast implants after the agency's 1992 decision, but disputes over the devices continued. The debate focused on two issues: the possibility that silicone could affect the immune system and the legal remedies for women claiming to be victims of implants. This epilogue sketches the scientific and legal events that have taken place since 1992. It also updates the stories of several of the major players in the breast implant controversy. For these groups, the claims and constructions made during that period have had an enduring significance. In the final pages, we revisit some of the questions first raised in the introduction.

After the FDA hearings and the attendant publicity about a possible link between implants and autoimmune disease, there was a flurry of publications about silicone in medical and technical journals. These articles presented evidence both for and against a causal association. Researchers convinced that silicone was harmful published review articles summarizing the laboratory studies and clinical reports that had been the subject of testimony. Like the witnesses who had argued against implants, these scientists concluded that silicone was neither inert nor nonreactive and suggested that it could have effects on the immune system.[1] Other reports proposed mechanisms by which those effects might work.[2] Skeptics drew different conclusions. Published statements from the ASPRS, the Department of Health in Great Britain, an independent advisory committee of Canadian physicians, and the American Medical Association all concluded that there was no evidence to connect silicone exposure to the development of autoimmune disease.[3] Both sides agreed, however, that more research—in particular, large-scale epidemiological studies—should be performed.

Beginning in late 1992, a series of laboratory and clinical studies seemed to bolster the claim that silicone could be immunogenic. One study reported that patients with an inflammatory reaction to internally implanted shunts fabricated of silicone rubber had developed silicone-specific antibodies;[4] another that women with implants who complained of connective tissue disease–like symptoms showed similar antinuclear antibody titres (a marker for autoimmune disease), as did patients with diagnosed connective tissue disease and no implants.[5] Studies of solid silicone joint replacements linked the material to localized inflammatory responses.[6] Lawyers examining the Dow Corning documents discovered a 1975 study showing that one of the low-molecular-weight silicone polymers contained in silicone gel could cause depressed immunity in mice.[7] The adjuvant hypothesis was supported by experimental work with rats, which demonstrated that injections of silicone gel increased the rats' immune response to foreign proteins.[8] A clinical report described a syndrome of "sclerodermalike esophageal disease" in children breast-fed by women with silicone breast implants.[9] As these reports emerged, defenders of implants disputed their validity, noting that many of the authors in fact earned the bulk of their incomes by testifying for plaintiffs in suits against implant manufacturers.

Results of the first two epidemiologic studies were released in the spring of 1994. These results contradicted assertions of a causal connection between silicone implants and certain types of autoimmune disease. The first, a case-control study performed by researchers at the University of Michigan, found no association between silicone gel breast implants and scleroderma.[10] A month later, the authors of a retrospective cohort study performed at the Mayo

Clinic found no association between implants and several autoimmune and connective tissue diseases.[11] In 1995, a study using data from the Harvard Nurses Health Study reported that there was no association between implants and connective tissue disease or the signs and symptoms of these diseases.[12] In 1996, however, another Harvard study found a 24 percent greater relative risk for self-reported connective tissue disease among women with implants when compared to a population without implants.[13]

Reactions to these publications reemphasized the gulf between antagonists in the implant controversy. Although the studies' authors were careful to point out their methodological limitations and possible sources of bias, the first three reports—the work done at Michigan, Mayo, and Harvard—were characterized by the popular press as definitive proof of implant safety.[14] Proponents of implants leaped upon this interpretation, promoting the epidemiologic work as vindication. The studies also were convincing to mainstream medical groups with less partisan stakes in the implant issue. In 1995, the American College of Rheumatology issued a consensus statement that reflected the lack of evidence for a causal association between implants and connective tissue disease. In 1997, the Practice Committee of the American Academy of Neurology did the same.

Those invested in the reality of silicone harm, however—silicone victims, their doctors and lawyers—contested the interpretation of safety. Their challenge to the validity of the epidemiologic research encompassed methodological critiques (mainly that the numbers of women studied were not large enough to give a definitive answer) but also introduced two new arguments: first, that any study funded by implant manufacturers was, on its face, invalid; and second, that the studies had failed to find a connection between implants and illness because the researchers had looked for classic connective tissue and autoimmune diseases, when silicone disease was a new and distinct entity with its own constellation of signs and symptoms.[15] Predictably, implant opponents seized upon the second Harvard study and its finding of a 24 percent increased risk of self-reported illness in women with implants. Implant proponents, however, focused on the absolute size of that risk (as well as the fact that it was based on self reports), emphasizing that it would account for only one extra case of connective tissue disease per three thousand women in the population.

The FDA took note of the epidemiologic reports but did not change its policy. In interviews Commissioner Kessler conceded that the new studies suggested that there was not any large risk of autoimmune disease from implants. He continued, however, to stress all that was still unknown about the devices—in particular, the rate of rupture.[16] In a 1996 publication, several FDA scientific staffers reviewed the currently available epidemiologic evidence and found that while "studies of scleroderma and other defined connective tissue diseases suggest that implant recipients have no substantially increased risk for these disorders . . . the epidemiologic literature is insufficient to rule out an association between breast implants and connective tissue disease–like syndrome." Despite continuing research, the authors concluded, "information is insufficient to adequately advise women who currently have or are seeking to obtain breast implants about the overall risk of these devices."[17]

Abroad, however, countries that initially had followed the lead of the FDA read the emerging evidence as reason to diverge from its policies. Spain, which had instituted a ban on implants following the FDA's 1992 moratorium, reversed the policy less than a year later.[18] Similarly, France, which had also followed the FDA in instituting a moratorium, in 1994 made implants available again under the provisos that all implants receive either French or European certificates of conformity (an indication that the devices met manufacturing standards) and that all patients be listed with a national registry for long-term follow-up.[19] In Britain, which had never restricted the use of implants, a 1994 report from a government-associated agency reviewed the available scientific evidence and exonerated implants almost entirely. (Dow Corning soon distributed the agency's report in the United States.)[20]

Research on silicone gel breast implants has continued at a furious pace. A recent Medline search of publications from 1996 to mid-1998 resulted in a list of citations that ran more than thirty pages. Their topics reflect the aspects of implants that historically have been problematic: gel bleed and gel migration; implant rupture; cancer detection; carcinogenesis (a 1997 update of the Deapen study concluded that implants seemed actually to protect women from breast cancer);[21] capsular contracture; case reports of odd syndromes in women with implants; psychological studies of women who seek the devices. Research on the connection between silicone implants and autoimmune disease also has continued. Published reports of laboratory and clinical studies reach conflicting conclusions (and draw a vociferous correspondence in the letters columns of these scientific journals). More epidemiologic studies have appeared, and—as a plethora of review articles (many written by plastic surgeons or published in plastic surgery journals) argue—none shows any association between implants and any form of classic or atypical autoimmune disease.[22] In July of 1998, after conducting a thorough review of the extant evidence, a panel of scientists appointed by the British government issued a report finding "no credible evidence that silicone elicits abnormal immune system reactions."[23] A week later, in the United States, the Institute of Medicine (IOM), at the request of the National Institutes of Health, convened a panel of medical researchers to review all of the scientific evidence. As this book went into production, the IOM panel issued its report, stating that it found no link between implants and systemic disease.[24]

The broader context for arguments over the scientific research was the highly legalized environment that surrounded the breast implant issue. From 1992 to 1994, women claiming damage from implants filed approximately ten thousand individual suits against the implant manufacturers.[25] For those cases that actually went to trial, outcomes were mixed. Manufacturers won several cases when juries found that implants had not caused the plaintiffs' injuries.[26] In other cases, however, judgments for the plaintiffs resulted in huge damage awards. For example, in the spring of 1994 three plaintiffs won $27.9 million in a suit against the 3M Company and General Electric (companies that had manufactured some implant components).[27]

Plaintiffs filed product liability cases against implant manufacturers in both state and federal courts. In 1992, a judge consolidated the seventy-eight cases pending in federal courts—including a class action suit that had been certified earlier in the year—into a multidistrict litigation (MDL) panel.[28] The consolidation allowed one group of lawyers to conduct one discovery proceeding, which could then be used in all subsequent federal cases. Instituted as a matter of efficiency, the MDL panel quickly came to provide a locus for much of the activity surrounding breast implant litigation. Soon after it was constituted, the panel began settlement negotiations with the manufacturers named in the suits. In September of 1993, the MDL lawyers and the manufacturers announced a proposed agreement in which the manufacturers would pay into a $4.75-billion settlement fund that would be distributed to women claiming damages. In return, the claimants would agree to drop all present and future legal proceedings.[29]

Negotiations to fine-tune the deal—dubbed the global settlement—continued for almost a year. In April 1994, United States District Court judge Sam C. Pointer of Alabama, who had been appointed to supervise the multidistrict litigation, gave preliminary approval to the terms of the settlement.[30] Soon afterward, legal notices in media outlets across the country notified women with implants about the possible settlement and urged them to write for more information.

The settlement structured by the agreement can be summarized as follows.[31] Women with implants were given a choice. They could join the settlement, which meant agreeing to the compensation terms worked out by the MDL panel and the manufacturers. If they chose to join, they would lose the right to file an individual suit. Women who did not wish to be part of the settlement had an affirmative duty to drop out—those who did not file a form with the MDL administrator stating their refusal to join the settlement would, by omission, be assumed to have "opted in" and would be bound by its terms.

The $4.75-billion settlement included several different funds: one for women with current disease, one for women who might develop such disease in the future, one for women who claimed other medical expenses, another for second-generation claimants, and another—limited to one-fourth of the total (or more than $1 billion)—for lawyers' fees. The settlement's compensation terms provided for a schedule of benefits structured by claimed condition, severity, and age at onset. The dollar amounts ranged, for example, from $1.4 million for a woman under thirty-six diagnosed with severe scleroderma, to $140,000 for a woman older than fifty-six diagnosed with mild atypical connective tissue disease. Claimants were required to provide medical documentation of their condition but were under no mandate to prove causation; that is, a woman need only show that she had been diagnosed at some point after receiving implants. The manufacturers reserved the right, depending on the number of claimants, to reduce the compensation amounts. If such a reduction were to occur, claimants would receive another opportunity to opt out.

The *New York Times* called the proposed settlement the "largest and most complex class action settlement in history."[32] Its size and complexity reflected the extent to which each side was attempting to hedge its bets; essentially, the settlement was a gamble over the caprice of the jury system. In deciding whether to opt in, claimants had to consider whether they stood to gain more by filing individual suits—likely to entail a long legal process with high costs and a chance of failure, but holding the possibility of a huge financial award—or by joining the global settlement, which would require no proof of causation, would guarantee a definite payment (although much less than they might expect in court), and would not subtract attorneys' fees from the settlement amount. The manufacturers, on the other hand, were gambling that the billions of dollars in settlement costs would result in a savings over the costs of fighting thousands of individual suits and losing some proportion of these suits. In agreeing to the terms of the settlement, the manufacturers were adamant that their capitulation was not an admission of wrongdoing. Rather, they claimed that the settlement simply represented their best business decision.

By the spring of 1995, 440,000 women had registered with the MDL claims administrator. Nearly 250,000 had filed claims for current disease compensation. Of these, the MDL administrator estimated that some twenty thousand claims would be approved quickly and that fifty thousand claims had minor deficiencies but had the potential to be approved. More than ten thousand claims had been filed by foreign women who had received implants produced by American manufacturers. In addition, approximately fifteen thousand women had opted out of the global settlement, thus reserving their right to sue individually.

In May 1995, Dow Corning filed for bankruptcy protection. In

August, Judge Pointer delayed his final ruling on the settlement in response to legal actions taken by the lawyers for foreign claimants, who were objecting to the smaller amounts allocated for their compensation. In the fall of that year, the settlement collapsed.

The collapse was due to a number of factors, including the Dow Corning bankruptcy and the epidemiological studies that had been appearing as the settlement developed. As the organizations of silicone victims had predicted early on,[33] the amount of funding proved insufficient for the number of claimants—a number that increased as the sure money of the global settlement became more appealing to plaintiffs who would be forced to make their cases in the face of contradictory scientific evidence. The lack of funding became more acute when Dow Corning's bankruptcy declaration effectively removed its $2-billion contribution from the settlement fund. The high number of opt-outs decreased the attractiveness of the settlement for the manufacturers; the economies they had sought in settling could no longer be assured. In addition, the new epidemiological evidence suggested to the manufacturers that they might not do so badly taking their chances in court.

In 1995, several of the manufacturers involved in the global settlement (but not Dow Corning) quietly reached a new settlement agreement (somewhere in the neighborhood of $3 billion) and began distributing money to their claimants. Because of Dow Corning's protected status, it was at this point legally untouchable. Plaintiffs' lawyers tried a new approach, claiming that Dow Chemical, one of the parent companies of Dow Corning, also bore legal responsibility for implants. While some state courts (e.g., New York) rejected the assertion of Dow Chemical's liability, others accepted the theory and juries began finding against the company.[34] Women who had opted out of the new settlement, and whose claims were against companies other than Dow Corning, continued to sue in state and federal courts. As before, the results of these cases were mixed, with some juries coming back with huge damage awards for the plaintiffs and others finding no evidence that implants had caused their problems.

The typical breast implant trial featured scientific testimony by opposing experts. Plaintiffs' experts, who spoke of silicone-specific antibodies and atypical connective tissue disease, presented plausible-sounding theories of harm. Defendants' experts, on the other hand, used the accumulating evidence from the epidemiologic studies, as well as critiques of their opponents' (un)scientific methods, to refute these theories. For jurors with no particular scientific expertise, decisions rested not so much on logic but on more volatile combinations of sympathy and anger. Following the 1993 Supreme Court decision in *Daubert v. Merrill Dow Pharmaceuticals* (a decision that gave judges a new responsibility to assess the validity of scientific evidence before allowing it in a trial),[35] judges assigned to breast implant litigation increasingly used this "gatekeeping" authority to prevent plaintiffs from even presenting evidence about theories of silicone disease and its causation. Similarly, Judge Pointer, who continued to oversee the MDL panel, appointed a group of nonpartisan scientific experts (i.e., scientists and physicians who had never testified in a breast implant trial or taken a public position on implants) to review the extant scientific evidence and "determine whether existing studies and research provide a scientific basis for the proposition that silicone causes or exacerbates the symptoms plaintiffs allege in their lawsuits."[36] In December of 1998, this expert review panel issued a report that concluded none of the scientific evidence supported a causal connection between implants and autoimmune disease. Reviews by other groups claiming disinterested status, however, argued that much of the extant scientific evidence was severely flawed.[37]

The continuing struggle over the science of implants constructed a new problem: a challenge to the idea of what constituted "valid" science. Proponents and opponents of implants each questioned the validity of the other's science. These challenges included disagreements over methodology and interpretation—the traditional areas over which science could be disputed. They also broadened the critique, raising challenges to the notion of who was qualified to perform research (the question of who could be certified as an expert), to the source of funding for research, and to the effect of intended audience on research findings. Validity was being reconceptualized, stretching from the technical to encompass the social. That the courts were relying on an ideal of scientific validity—as in the *Daubert* decision—to solve the "junk science" problem seemed increasingly ironic.

For the players who had been front and center during the FDA controversy, the agency's 1992 decision ended one act but began another. In November of 1992, Dow Corning hired former U.S. attorney general Griffin Bell to conduct an internal investigation of the company's "stewardship" of the implant business. Although Bell's report was never released in its entirety, Dow Corning claimed that it was largely exculpatory. In 1994, several Dow Corning employees published an article in the medical journal *Arthritis and Rheumatism* that characterized the breast implant controversy as "a contentious issue involving a complex and dynamic interplay of science, culture, litigation, economics, innovation, regulation, and the news media," [38] which had produced "an eclectic mix of opinions, a Babel of different positions."[39] The authors argued that throughout the controversy "industry was trying to determine what data were needed in order to be responsive to the evolving regulatory requirements, and simultaneously arguing that the public was best served by a regulatory climate that supported continued opportunities for innovation." However, "the lawyers had identified an opportunity."[40]

Seeing themselves as the target of this "opportunity," in May of 1995 the company retreated into bankruptcy protection. As noted earlier, this status insulated Dow Corning from new liability suits. The company still had to deal with the thousands of already existing lawsuits, however. (Estimates placed the total number of suits against Dow Corning at approximately 170,000.) The court-appointed bankruptcy reorganization panel tried to find ways to address these claims. In 1996, Dow Corning proposed that the plaintiffs be allowed to take their claims to court, but that the individual trials be preceded by one large "science trial" that would address all of the issues of causation. (Clearly the company believed that a favorable outcome in the science trial would protect them from huge damage awards in the individual trials.) This idea failed when groups of silicone victims demanded that they, as well as the scientific experts, be given an opportunity to testify at such a trial. In 1997, Dow Corning floated a settlement offer of some $2.4 billion. Reaction from plaintiffs and their attorneys suggested that the sum was too low. In July of 1998, Dow Corning made another, formal, offer, this time for $3.2 billion.

Throughout the controversy and its aftermath, Dow Corning and the other manufacturers kept their economic interests primary. When it looked as though implants could be approved by the FDA, industry was willing to make the financial investment necessary to support research and testing. Once it became clear that the implant business would become one of diminishing returns—as profits were eaten up by the costs of clinical trials and litigation—however, most of the manufacturers changed their minds, seeing in withdrawal the most cost-effective action.

Concurrent with its withdrawal from the market, Dow Corning implemented an image rehabilitation strategy. This effort began with an attempt to portray itself as a David of industry—"founded 50 years ago in a small midwestern town, Dow Corning has prided itself on being a producer of 'good' materials, the silicones"[41]—facing a Goliath of greedy plaintiffs' attorneys. On May 10, 1995 (less than a week before it filed for bankruptcy protection and precipitated the collapse of the global settlement), the company placed a full-page advertisement in major newspapers across the country. The ad began, "Here's what some people don't want you to know about breast implants," and went on to emphasize the findings of the epidemiological studies, the decisions made by governments abroad to end restrictions on the devices, the ubiquity of silicone, and the financial interests and political power of trial lawyers. (Readers who wrote or called in response to the ad received a packet of information about each of these topics.) Dow Corning used its financial support for many of the epidemiological studies to ensure good publicity: when these studies appeared to exonerate implants, the company actively promoted them in the popular press. For example, Dow Corning released the results of the Michigan study to the press before the article about it was submitted to a journal.[42]

The ASPRS, too, emerged from the breast implant controversy and immediately turned its attention to image rehabilitation. Like Dow Corning, the organization appropriated the stance of victim. Soon after the FDA's 1992 decision, a number of plastic surgeons were interviewed for a news article in *JAMA*. The author of the piece located the genesis of the implant problem in "inappropriate" patient demand, claiming that a "surgical procedure introduced thirty years ago with the promise of restoring wholeness to the relatively few women who lost a breast to therapeutic mastectomy . . . was co-opted by hundreds of thousands of their sisters as a seemingly easy means of gaining the buxomness Mother Nature may have denied their own two healthy breasts."[43] Caught between their overly demanding patients and the FDA, plastic surgeons found themselves like "wildebeests" surrounded by predatory hyenas—plaintiffs' attorneys.[44]

In his final speech to the organization, ASPRS president Norman Cole (who left office soon after the breast implant decision) faulted the FDA, the Bush administration, and, especially, the AMA for failing to support the plastic surgeons during the controversy.[45] In a later interview, Cole blamed the FDA for capitulating to pressure from special interest groups—in particular, Public Citizen, which he claimed received most of its funding from trial lawyers. Rather bizarrely, he speculated whether "this whole thing was a way to pander to the religious right." "As a specialty," Cole concluded, "we were naive in assuming that truth, science, and reason would decide the silicone crisis. . . . This was and is being decided by the politics of a special interest group. This wasn't a campaign against silicone; it was a war against breast implants."[46]

Perhaps most devastating to the plastic surgeons, however, was the perceived betrayal by the manufacturers. Cole charged that the implant industry had deceived the plastic surgeons at every turn; for example, the manufacturers had misled them as to the likely ease of gaining FDA approval and had blindsided plastic surgeons by concealing the Hopkins litigation until the verdict hit the newspapers. Other plastic surgeons called "the irreparable severing of trust that previously existed between doctors and device manufacturers" one of the "most destructive" repercussions of the implant controversy.[47]

The extent of the split between the plastic surgeons and the implant manufacturers was revealed in the legal wranglings over implants. In court, each side blamed the other. Plastic surgeons defending themselves against malpractice suits blamed the manufacturers for failing to disclose the risks of implants. The manufacturers used a "learned intermediary" defense to contest their liability, claiming that the duty to warn the patient resided with the treating physician and not the product manufacturer.[48] The provisions of the global settlement hardened the antagonism. The settlement resolved the liability claims against the manufacturers, but it offered no protection for individual plastic surgeons. (A woman who opted in to the settlement would be enjoined from suing the manufacturers but could still sue her plastic surgeon.) In an amicus curiae brief to Judge Pointer, the ASPRS objected to these terms. Privately, ASPRS lawyers began negotiations with the MDL panel lawyers for an "in-kind services" agreement in which participating surgeons would be released from all liability charges from class members (i.e., women who had opted in to the settlement) by agreeing to perform a certain number of gratis surgical procedures.[49] Eventually Judge Pointer rejected the proposal, noting that he had "reservations about a program that would require a woman to opt out of the settlement . . . in order to bring an action against a participating surgeon."[50] When the settlement collapsed, the matter became moot.

In their tattered relationship with the manufacturers, the plastic surgeons once again represented themselves as victims, bemoaning their own helplessness: "Never again should medicine be willing to use a product from industry without being able to see the research data. We never would have accepted those answers from our own colleagues, but by law we had to depend upon the integrity of the manufacturer."[51]

The ASPRS announced that the money left over from its "special assessment"—the money levied to fight for implants—would be used for "government relations and socioeconomic and public education efforts," including the establishment of a "'spokesperson network,' an image campaign, and a new campaign of outreach to primary care physicians."[52] In addition, the ASPRS allocated about half a million dollars to long-term studies of implants.[53] Looking to the future, the ASPRS noted that while the "crisis" had been difficult for the plastic surgeons, it had "proved without a doubt that the ASPRS had come of age. It could take its place proudly at the national table."[54] (The statement proved prophetic. In the late 1990s, for example, the ASPRS was one of the moving forces behind a congressional bill requiring insurers to cover the costs of reconstructive breast surgery.)[55]

The movement of silicone victims grew stronger in the wake of the FDA decision. New support groups were formed and began publishing newsletters and, taking advantage of new technology, went on-line. Representatives of established groups, like CTN, made frequent media appearances. (And then used these media appearances as proof of their own legitimacy.) These groups focused their efforts on education and research, advocacy and support, and provided public forums for personal narratives of suffering. All of their activities were premised on the danger of silicone implants.

As the global settlement took shape, support groups devoted themselves to describing its terms and advising women of their rights and responsibilities. (The rhetoric of choice cropped up as a way to describe women's legal options.) Increasingly, the pages of their newsletters were filled with advertisements from financial services and investment firms. When the MDL administrator released reports of the number of women who had registered for the settlement, these numbers were quickly seized upon by victims' groups and attorneys as proof of their claim of harm. (*Ms.* magazine took the same opportunity. Its March-April 1996 issue had a cover story on "the ongoing cover-up" with the teaser "Can 400,000 women *really* be wrong about the dangers of silicone?") In 1995, the claims of the silicone victims were bolstered by the publication of *Informed Consent*, a book-length exposé of breast implants that told the story of a Dow Corning executive whose wife became ill after receiving the devices.[56]

As the results of the epidemiologic studies gained general acceptance, the victims' groups lost some of their credibility. It became more common for observers to express skepticism about the victims' claims of illness—to blame financial self-interest or psychiatric disturbance and not silicone for their complaints. In the face of such disapprobation, the victims' groups became even more important for individual women who believed that their implants had made them sick. Despite the growing skepticism about their claims, the victims' groups held their place among the stakeholders in the implant issue. The media continued to quote them; they remained a presence at implant-related public events. In July of 1998, a coalition of silicone victims' groups asserted their political power by filing suit against the ASPRS and the FDA, claiming that they had failed to warn women of the dangers of implants. The suit called for an immediate and total prohibition on all silicone breast implants.[57]

While the controversy raged, the plight of women who had sought implants for reconstruction received much sympathetic publicity. Implant advocates, many of them women who had received the devices for reconstruction after mastectomy, consistently portrayed implants as a problem of breast cancer. In tying implants to the interests of the breast cancer lobby, the advocates hoped to insulate implants from restrictive regulation. (In so doing, however, they opened the door to the FDA's policy distinction between reconstruction and augmentation, a distinction that would reduce the market for implants by about 75 percent.) It was during this period that breast cancer advocates seeking to promote a broader agenda became increasingly visible. In 1991, several breast cancer advocacy groups joined together to form the National Breast Cancer Coalition. The coalition's goal was to "challenge" the national breast cancer research agenda: that is, both to demand more funding for breast cancer research and to secure for breast cancer advocacy groups some influence over research priorities. Using extremely creative strategizing, in 1992 the coalition was able to claim its first large success: a $200-million allocation for breast cancer research in the budget of the Department of Defense. In 1993, several federal agencies came together with the coalition and other breast cancer advocates for a conference to establish a "National Action Plan on Breast Cancer." The plan laid out priorities in health care, research, and policy, priorities that have begun to be implemented as Congress has increased federal spending on breast cancer each year since 1993.[58] In the process of their lobbying efforts, the breast cancer advocates became extremely influential in Washington. Politicians sought their endorsement for most major health care legislation. The bill requiring managed care organizations to provide reimbursement for reconstructive surgery after mastectomy obtained almost instant legitimacy on Capitol Hill because it was backed by several of the most powerful breast cancer advocacy groups.

The saga of implants continued in the media. Before 1988, the press largely accepted the claim that implants were safe. Most stories took a "gee whiz" angle, describing, for example, the "miracle" of reconstruction for women whose lives had been devastated by breast cancer. In these articles, plastic surgeons were portrayed as benevolent experts. The coverage turned more skeptical during the period from 1988 to 1990, as more of the silicone victims' stories received attention.

After a frenzy of stories in 1991–92—the height of the controversy—that constructed the problem of implants as one of harm, the press was equally quick to promote the epidemiologic studies as proof of safety. By 1995–96 much mainstream coverage of the implant story had turned reflective. The new "take" on the implant issue was exemplified by a *60 Minutes* segment and a *Frontline* documentary broadcast on PBS.[59] Both reconstructed the problem of implants as one of societal overreaction fueled by the misuse of science in the courtroom, and interpreted the lesson of implants as a need for tort reform. The ambiguity portrayed was not the safety of silicone but how opponents of breast implants had managed to carry the day without any valid evidence.

The FDA emerged from the controversy in a strong assertion of its authority, but with lingering questions as to its broader legitimacy. The agency's fortunes shifted with the election of a Democratic administration, whose pledge to reinvent government stopped somewhat short of eviscerating all federal regulatory agencies. The FDA continued to be bombarded with criticisms and challenges, however. The success of AIDS activists inspired lobbying groups for other diseases to demand streamlined drug approval processes for themselves. Periodic reports of harm linked to FDA-approved drugs or devices called into question the competence and independence of the agency. Industry decried the strangling effect of government regulation. Talk of FDA reform continued on Capitol Hill. After the election of a Republican Congress in the fall of 1994, these reform efforts intensified. Bills introduced in the House and Senate in 1996 echoed the plan devised by Quayle's Council on Competitiveness, calling for shorter review times for drugs and devices, privatization of the review process, approval reciprocity with the United Kingdom and the European Union, and the increased promotion of off-label uses for drugs and devices. (Interestingly, as reform was being debated by the Republication

Congress, the FDA—under the leadership of David Kessler—was asserting its authority once again, this time by claiming jurisdiction over tobacco.) The reform legislation that eventually passed in 1997 stopped short of some of these recommendations but incorporated many of the salient ideas. [60] In mandating expanded access to investigational drugs and devices, introducing policies that called for increased collaboration between the FDA and other federal agencies involved in science and health, and structuring opportunities for a private review process for some drugs and devices, Congress was clearly seeking to broaden the agency's accountability to groups and institutions outside the traditional FDA-industry dyad.

And what of the implant itself? After the FDA's decision to limit the availability of silicone gel breast implants, saline implants became the only option for women seeking augmentation or for reconstruction patients who did not wish to be part of a clinical trial. Perhaps because of the publicity about silicone gel implants, or because of questions that arose about the safety of saline-filled implants,[61] the number of augmentation surgeries fell precipitously after the FDA's decision. In 1992, there were just over 30,000 augmentations performed (compared to well over 120,000 in 1990). As time has passed since the immediacy of the controversy, however, and as recent highly publicized studies have appeared to exonerate implants, that number has steadily increased. In 1997, there were once again over 120,000 augmentations performed.[62] Recent press reports suggest that an increasing number of augmentations are being performed using silicone gel implants purchased in the black market.[63]

Plastic surgeons are once again taking up the cycle of technological innovation. A 1996 article in *Plastic and Reconstructive Surgery* described more than nine alternative implant filler materials that were in some stage of development.[64] In echoes of the past, articles in the popular press have described the resurgent use of autogenous fat for augmentation (one technique uses the acronym BAMBI, for Breast Augmentation Mammoplasty by Injection).[65] And, most recently, the biotechnology company Reprogenesis announced plans to begin testing genetic engineering techniques that may eventually allow the firm to grow breasts from a woman's own cells.[66]

To do breast self-examination, one moves from the center of the breast outward, probing in ever widening circles. This is an apt metaphor for the story that has been told here. We began at the center, with the breast implant. Our examination of the history of breast implant technology quickly moved outward to look at the "concepts . . . [and] political, social and economic structures"[67] bound up with the device and then to probe the broader meanings that various actors sought to enforce through their claims about implants.[68]

As noted in the introduction, there are two ways of thinking about the significance of technology. In the first, technology is conceptualized as a symbol, a reflection of the society that created it. This approach lies behind most of the extant research on breast implants and cosmetic surgery, much of which has focused on the question of why women desire breast implants but has ignored the actual breast implants. The alternative approach changes the perspective by conceptualizing technology as an instrument that—while reflective—is also used to shape society. Using the latter approach in this book has shifted the focus of inquiry away from questions of individual desire (and the paradoxes of choice) and toward an examination of how, and by whom, the public meanings embedded in a specific technology are constructed and then how, and on whom, these meanings come to have broad effects.

During the hundred years that passed between Czerny's case report and the FDA's regulatory decision, the meaning of the breast implant underwent many revisions. In its nascent stage—the era of paraffin injections and autogenous fat transplantations—implant technology stood for some sense of vague dissatisfaction. For women, that dissatisfaction focused on the body; for plastic surgeons, on the sense that some potential was not yet fully grasped. With the advent of synthetic sponge materials after the Second World War, the meaning of breast implants began to come into sharper focus. The improved technology allowed a broad medicalization of aesthetic problems. The vague dissatisfaction that plastic surgeons had attributed to their patients was pathologized. The desire for breast implants became a disease marked by both anatomic and psychological signs and symptoms. As use increased, tension grew within the profession over the possible dangers of the shift to artificial materials.

Technological advances—in particular, the introduction of silicone—prompted reconsideration of earlier technologies, of the use of artificial materials, and of patients. The historiography of breast implants took on a teleological slant—all had been building to the introduction of this, the ultimate technology. The tension between the natural and the artificial was dispersed through a redefinition of the normal: silicone was the artificial technology that surpassed the natural, thus allowing its use to seem just as natural. The pathology of desire was revised away. The desire for breast implants was reconceptualized as a sign of health and liberation. Plastic surgeons sought legitimization for implants through their special empathy—characterized, in part, by the validation of vanity—and, less successfully, through an attempt to promote implants as a preventive treatment for breast cancer, one of the most intractable of women's health problems.

With the dawning realization of the possibility of implant-caused harm, the meaning of the breast implant once again was revised. For an increasingly visible group of victims, the implant was danger incarnate, the cause of a new entity called silicone disease. In promoting this meaning, victims sought broad recognition for the perfidy that had harmed them and some remedy by which they could be compensated. Groups of implant proponents—satisfied recipients, plastic surgeons, representatives of the implant industry—challenged this attempt to attach such negative meanings to implants, trying instead to construct implants as emblems of individual autonomy and choice.

As implants moved into the regulatory arena, the scope of the argument broadened. Antagonists in the implant controversy sought to control the outcome of the regulatory process by controlling the public meaning of the technology. Implants came to stand for multiple controversies, including those over the authority of the FDA, the definition of scientific validity, and the essentially social, rather than technical, nature of risk assessment. The FDA's 1992 resolution of the implant matter did not fully decide these issues but shifted them into other arenas—where, needless to say, the construction of meaning continues.

The instrumentality of implants includes their private and their public meanings. Privately, the power of implants encompasses individual desire, the "choice" of a particular body. Publicly, their power extends over a variety of controversies (public dramas) and disputes. Both privately and publicly, this power is the result of human action—the claims and arguments that serve to construct meaning. The problem of implants is a problem of clashing meanings. The solution lies not in the science of epidemiological research but in the trans-science of ideology.

At the beginning of this book I offered two sets of images, intimating that my task would be to interpret and make sense of the history that linked them. In the first, four highly intelligent and poised women advocated for their right to choose implant technology. In the second, some anonymous and naked female body, disempowered by the framing of the photograph (cropped headless and handless), stood to demonstrate that technology. The juxtaposition of these images epitomized many of the ironies of implants.

One irony is the contrast of desire and need. If the four advocates represent the former, the naked body represents the latter. Historically, need was constructed in order to legitimize desire. In the portraits of four women, however, there is a subtle reversal, and desire is used as a way to justify need. Another irony is voice. The headless body is forever silent; the handless body gestureless. The four advocates, in contrast, are highly articulate. It is their words that are quoted. Their power resides in their voices. But for whom are they speaking? It is hard to ignore that their statements echo the propaganda of the ASPRS, that the ASPRS paid for the ad, that these women embody the organization's crisis-driven strategy of recruiting attractive and articulate patients as activists. Nor can we ignore that the headless, handless woman did have a voice, a will. The movie girl may not have spoken, but she did assent. (And she did blow that kiss.) A third irony lies in the visualization of the implant in the images. While in the before-and-after photographs implants are primary, the focus of the camera and the eye, the cropping of the ASPRS advertisement serves to elide their existence. The elision serves the political goals of the advertisement. It suggests that the problem of implants has nothing to do with implants, but with overzealous government regulators, and it emphasizes the plastic surgeons' purported concern for the whole woman. But without need and without the implant, why is there a need for implants? Absurdity abounds. Is it not ridiculous that this "ultimate symbol of femininity" has become both a product of industry and an obsession of government, two masculine institutions? That implants can be, simultaneously, the subject of so much serious discussion and a dirty joke? That cleavage has wrought so much . . . cleavage? When it comes to breast implants, it's all just a little bit over the top.

Notes

INTRODUCTION

1. Emily Martin, *The Woman in the Body: A Cultural Analysis of Reproduction* (Boston: Beacon Press, 1987).
2. Susan Bell, "A New Model of Medical Technology Development: A Case Study of DES," *Research in the Sociology of Health Care* 4 (1986): 3.
3. Donna J. Haraway, "A Cyborg Manifesto: Science, Technology, and Socialist-Feminism in the Late Twentieth Century," in *Simians, Cyborgs, and Women: The Reinvention of Nature*, (New York: Routledge, 1991), 164.
4. In this summary I am by necessity simplifying what is a much more complex and subtle argument about the concept of "discipline." That argument is expounded in Michel Foucault, *Discipline and Punish: The Birth of the Prison*, translated by Alan Sheridan (New York: Pantheon Books, 1977) and Susan Bordo, "'Material Girl': The Effacements of Postmodern Culture," *Michigan Quarterly Review* 29, 4, (1990): 653–657. Bordo, for example, describes "a dynamic of non-centralized forces, its dominant forms attaining their hegemony, not from magisterial design or decree, but through multiple 'processes, of different origin and scattered location' regulating and normalizing the most intimate and minute elements of the construction of time, space, desire, embodiment" (666). To read how versions of this argument have been used to explicate the phenomenon of cosmetic surgery, see Carole Spitzak, "The Confessional Mirror: Plastic Images for Surgery," *Canadian Journal of Political and Social Theory* 12, 1–2 (1988): 38–50; Diana Dull and Candace West, "Accounting for Cosmetic Surgery: The Accomplishment of Gender," *Social Problems* 38, 1 (1991): 54–70; and Anne Balsamo, "On the Cutting Edge: Cosmetic Surgery and the Technological Production of the Gendered Body," *Camera Obscura* 28 (1993): 207–237. For application to breast implants, see Linda Coco, "Silicone Breast Implants in America: A Choice of the 'Official Breast'?" in *Essays on Controlling Processes*, edited by Laura Nader (Berkeley: Kroeber Anthropological Society, 1994), 103–132; Nancy Datan, "Illness and Imagery: Feminist Cognition, Socialization, and Gender Identity," in *Gender and Thought: Psychological Perspectives*, edited by Mary Crawford and Margaret Gentry (New York: Springer-Verlag, 1989), 175–187; and Iris Marion Young, "Breasted Experience: The Look and the Feeling," in *Throwing Like a Girl and Other Essays in Feminist Philosophy and Social Theory* (Bloomington: Indiana University Press, 1990), 189–209.

5. Kathryn Pauly Morgan, "Women and the Knife: Cosmetic Surgery and the Colonization of Women's Bodies," *Hypatia* 6, 3 (1991): 25–53.
6. For example, Lisa S. Parker writes that "feminists generally believe that women's decisions to seek breast augmentation cannot constitute competent, informed choices, but necessarily reflect 'false consciousness,' resulting from the overwhelming influence of a society that prizes large breasts" (70–71). She argues that the "existence of these social pressures, attitudes, and institutions does not . . . negate the right of women to control their bodies, nor does it render them incapable of this self-determination. Cosmetic Surgery . . . is not something that women . . . are coerced to undergo" (72). Lisa S. Parker, "Social Justice, Federal Paternalism, and Feminism: Breast Implants in the Cultural Context of Female Beauty," *Kennedy Institute of Ethics Journal* 3, 1 (1993): 57–76.
7. For one such study, see Kathy Davis, *Reshaping the Female Body: The Dilemma of Cosmetic Surgery* (New York: Routledge, 1995).
8. Irving Zola quoted in Peter Conrad, "Medicalization and Social Control," *Annual Review of Sociology* 18 (1992): 210. My understanding of medicalization follows the model developed by Peter Conrad and Joseph W. Schneider in *Deviance and Medicalization: From Badness to Sickness* (Philadelphia: Temple University Press, 1992), in which medicalization is seen as a claims-making process. As Catherine Kohler Riessman argues in her examination of the medicalization of women's health—"Women and Medicalization: A New Perspective," *Social Policy* 14 (1988): 3–18—for such claims to be successful they must serve the interests both of the medical establishment and of the broader society, including those groups that are the objects of the claims.
9. Joyce Maynard, "After the Silicone," *Self*, September 1993; 196–198, 213.
10. Ibid., 196–197.
11. In *The Sociological Imagination* (New York: Oxford University Press, 1959), C. Wright Mills identifies the "sociological imagination" as that quality which "enables its possessor to understand the larger historical scene in terms of its meaning for the inner life and the external careers of individuals" (5).
12. Joseph R. Gusfield, *The Culture of Public Problems: Drinking-Driving and the Symbolic Order* (Chicago: University of Chicago Press, 1981). Gusfield offers a model of the construction of public problems that is based on the concepts of "ownership" and "responsibility." Ownership is "the ability to create and influence the public definition of a problem" (10), including the ability to assign responsibility for the problem. Gusfield describes two types of responsibility: causal and political. The former refers to the "causal explanation of events," whereas the latter "looks to the person or office charged with controlling a situation or solving a problem" (13). "The structure of public problems," Gusfield writes, "is then an arena of conflict in which a set of groups and institutions . . . compete and struggle over ownership and disownership, the acceptance of causal theories, and the fixation of responsibility" (15).
13. For a treatment of silicone breast implants as an exemplar of these critiques, see my "Critical Perspectives on the Making of Women's Health Policy: The Case of Silicone Breast Implants in the United States," in *Cultural Perspectives on Reproductive Health*, edited by Carla M. Obermeyer (forthcoming from Oxford University Press).
14. My particular influences include the following: Herbert Blumer, *Symbolic Interactionism: Perspective and Method* (Berkeley: University of California Press, 1969); Peter L. Berger and Thomas Luckmann, *The Social Construction of Reality: A Treatise in the Sociology of Knowledge* (New York: Anchor Books, 1966); Malcolm Spector and John I. Kitsuse, *Constructing Social Problems* (New York: Aldine de Gruyter, 1987); Stephen Hilgartner and Charles L. Bosk, "The Rise and Fall of Social Problems: A Public Arenas Model," *American Journal of Sociology* 94, 1 (1988): 53–78; as well as Gusfield's *The Culture of Public Problems*.

CHAPTER 1 STAGING BABEL: A CHRONICLE OF THE CONTROVERSY

1. These terms appear in, respectively: unpublished documents produced by the ASPRS; *Kirkus Review* excerpt appearing on the book jacket of John A. Byrne, *Informed Consent* (New York: McGraw-Hill, 1996); Byrne, *Informed Consent*; Henry Jenny, *Silicone-Gate: Exposing the Breast Implant Scandal* (Siloam Springs, Ark.: 1994); Nancy Bruning, *Breast Implants: Everything You Need to Know* (Alameda, Calif.: Hunter House, 1992).

2. Joseph R. Gusfield, *The Culture of Public Problems: Drinking-Driving and the Symbolic Order* (Chicago: University of Chicago Press, 1981).
3. Congressman Ted Weiss quoted in *Is the FDA Protecting Patients from the Dangers of Silicone Breast Implants? Hearing before the Human Resources and Intergovernmental Relations Subcommittee of the Committee on Government Operations*, House of Representatives, 101st Cong., 2d. sess., December 18, 1990, 2. Hereafter cited as *Weiss Hearing*.
4. The 1976 legislation required the classification of all devices into one of three regulatory categories. Class I, called "General Controls," allowed the manufacture and sale of devices for which general regulations specified in the 1938 Food, Drug, and Cosmetic Act were "sufficient to provide reasonable assurance of the safety and effectiveness of the devices." These regulations included bars against adulteration and misbranding of products and procedures for repair or replacement of defective products, as well as some record keeping requirements. Placement in Class II, "Performance Standards," was indicated when the general controls of Class I were deemed "insufficient to provide reasonable assurance of safety and effectiveness." Manufacture and sale of Class II devices would require a written standard that specified "provisions to provide reasonable assurance of . . . safe and effective performance," including "provisions respecting the construction, components, ingredients, and properties of the device . . . provisions for testing . . . provisions for the measurement of the performance characteristics of the device, provision [that the results of such testing be in conformity with] the portions of the standard for which the test or tests are required," provisions prescribing the form and content of labeling requirements, and a provision directing restrictions on sale and distribution if general conditions were not met. Class III, the strictest standard, was reserved for those devices for which there was "insufficient information" to indicate that placement in Class I or Class II could provide sufficient protection to the public. All quotations come from *Medical Device Amendments of 1976*, Public Law 94–295, 94th Cong., May 28, 1976.
5. Testimony of Rosemary Locke in *Weiss Hearing*, 11.
6. Testimony of Robert Rylee in ibid., 187.
7. "Analysis of Dow Corning Data Regarding Carcinogenicity of Silicone Gels," internal FDA memorandum from Mel Stratmeyer, chief, Health Sciences Branch. August 9, 1988. Reproduced in *Weiss Hearing*, 144–152.
8. Testimony of Robert Sheridan in ibid., 128.
9. The Public Citizen Health Research Group was founded in 1971. The organization's mission statement indicated that it intended "to fight for the public's health, and to give consumers more control over decisions that affect their health." The information about Public Citizen's role in leaking the rat study comes from Byrne, *Informed Consent* and the testimony of Norman Anderson in *Weiss Hearing*, 31.
10. The panel's recommendations are described in an FDA internal memorandum from the director of the Center for Devices and Radiological Health to the commissioner. The document is dated December 1, 1988. It is reproduced in *Weiss Hearing*, 171–175.
11. Testimony of Norman Anderson in *Weiss Hearing*, 30–39.
12. Ted Weiss in ibid., 27.
13. Ibid, 186.
14. Philip J. Hilts, "U.S. to Begin Regulation of Breast Implants in '91," *New York Times*, December 19, 1990.
15. Mary H. McGrath and Boyd R. Burkhardt, "The Safety and Efficacy of Breast Implants for Augmentation Mammaplasty," *Plastic and Reconstructive Surgery* 74, 4 (1984): 550–560.
16. For example, Louis H. Winer et al., "Tissue Reactions to Injected Silicone Liquids," *Archives of Dermatology* 90 (December 1964): 588–593; W. St. C. Symmers, "Silicone Mastitis in 'Topless' Waitresses and Some Other Varieties of Foreign-Body Mastitis," *British Medical Journal* 3 (July 1968): 19–22; Jerome S. Nosanchuk, "Silicone Granuloma in Breast," *Archives of Surgery* 97 (October 1968): 583–585; C. Hal Chaplin, "Loss of Both Breasts from Injection of Silicone (with Additive)," *Plastic and Reconstructive Surgery* 44, 5 (1969): 447–450; Khoo Boo-Chai, "The Complications of Augmentation Mammaplasty by Silicone Injection," *British Journal of Plastic Surgery* 22, 3 (1969): 281–285.

17. For example, K. Miyoshi et al., "Hypergammaglobulinemia by Prolonged Adjuvanticity in Man. Disorders Developed after Augmentation Mammaplasty," *Japan Medical Journal* 2122 (1964): 9–14; Seiichi Ohmori and Takashi Hirayama, "Is It Possible to Incur Adjuvant Disease in a Human Being by the Injection of Silicone?" *Japan Journal of Plastic and Reconstructive Surgery* 13 (1970): 162–166; K. Yoshida, "Post Mammaplasty Disorder as an Adjuvant Disease of Man," *Shikoku Acta Medica* 29 (1973): 318–332; Yasuo Kumagai, Chiyuki Abe, and Yuichi Shiokawa, "Scleroderma after Cosmetic Surgery," *Arthritis and Rheumatism* 22, 5 (1979): 532–537.
18. For example, Sheryl A. van Nunen, Paul A. Gatenby, and Antony Basten, "Post-Mammaplasty Connective Tissue Disease," *Arthritis and Rheumatism* 25, 6 (1982): 694–697; Harry Spiera, "Scleroderma after Silicone Augmentation Mammoplasty," *JAMA* 260, 2 (1988): 236–238.
19. For example, A. Vargas, "Shedding of Silicone Particles from Inflated Breast Implants" (letter to the editor), *Plastic and Reconstructive Surgery* 64, 2 (1979): 252–253; Nir Kossovsky et al., "Analysis of Surface Morphology of Recovered Silicone Mammary Prostheses," *Plastic and Reconstructive Surgery* 71, 6 (1983): 795–802.
20. Sybil Niden Goldrich, "Restoration Drama," *Ms.*, June 1988, 20–21.
21. Ibid, 20.
22. Ibid.
23. In *Informed Consent* (New York: McGraw-Hill, 1996), John A. Byrne describes this as Dow Corning's attitude in the years before the controversy broke.
24. The television program is described in ibid. A brief excerpt was shown in a PBS documentary about implants that aired on *Frontline* on February 27, 1996.
25. Marsha L. Vanderford and David H. Smith, *The Silicone Breast Implant Story: Communication and Uncertainty* (Mahwah, N.J.: Lawrence Erlbaum Associates, 1996).
26. Byrne, *Informed Consent*, 93.
27. Ibid., 93–107; Vedder, Price, Kaufman, and Kammholz, "Summary of Significant Decisions," unpublished document included in the ASPRS Breast Implant Litigation Handbook (also unpublished).
28. This review encompassed all articles about breast implants published in several widely read women's magazines (including *Harper's Bazaar*, *Glamour*, *Mademoiselle*, and *Vogue*) between 1950 and 1990. Articles were chosen based on titles listed in the *Reader's Guide to Periodical Literature* for the relevant years. Articles that did not include the word "breast" or "breast implant" in the title were not reviewed.
29. Letter from Norman Cole to ASPRS membership, October 4, 1991.
30. Letter from ASPRS leadership to organization members, fall 1991.
31. ASPRS letter, October 4, 1991.
32. "Media Tip," *Breast Implant Bulletin*, ASPRS publication, October 25, 1991.
33. "ASPRS to Use Ads to Communicate Directly with Women . . . Invites Regional Participation," *Breast Implant Bulletin*, ASPRS publication, October 25, 1991.
34. "Tips on Getting Involved," *Breast Implant Update*, ASPRS publication, November 1991.
35. "Update on Issues Surrounding Breast Implants," memorandum from ASPRS Breast Implant Crisis Task Force to ASPRS members, fall 1991 (unpublished document).
36. "Congress Listens to Plastic Surgeons, Patients," *Breast Implant Update*, ASPRS publication, November 1991. See also "Congress Asked to Help Preserve Women's Right to Choose Breast Implants," ASPRS news release, October 8, 1991.
37. Sheila Kaplan and Jonathan Grone, "'A Woman's Choice?' Guess Again, This Push Is for Breast Implants," *Legal Times*, November 11, 1991.
38. Christopher Drew and Michael Tackett, "Access Equals Clout: The Blitzing of FDA," *Chicago Tribune*, December 8, 1992.
39. Ibid.
40. Ibid.; Kaplan and Grone, "'A Woman's Choice?'"
41. FDA Talk Paper, August 22, 1991.
42. "PMA Applications for Silicone Breast Implants," memorandum from Diana Zuckerman to Congressman Ted Weiss, September 12, 1991.
43. Ibid.
44. FDA staffer Paul Tilton in transcript of the FDA's General and Plastic Surgery Devices Panel Meeting, November 12–14, 1991. Hereafter cited as November Meeting.
45. FDA commissioner David Kessler in ibid.

46. Vicky Castelberry (founder of Bosom Buddies) in ibid.
47. Sheila Profiter in ibid.
48. Mary Follin in ibid.
49. Sheila Profiter in ibid.
50. Evelyn Johnson in ibid.
51. Lois Green in ibid.
52. Angie Brown in ibid.
53. This movement was described in Sharon Batt, *Patient No More: The Politics of Breast Cancer* (Prince Edward Island: gynergy, 1994).
54. Janet van Winkle in November Meeting.
55. Sybil Niden Goldrich (founder of CTN) in ibid.
56. Cindy Pearson (National Women's Health Network) in ibid.
57. Lynda Roth (Coalition of Silicone Survivors) in ibid.
58. Cathy Price in ibid.
59. FDA staffer Joseph Levitt in ibid.
60. Vivian Snyder in ibid.
61. Rita Freedman in ibid.
62. Rosemary Locke in ibid.
63. James Potchen in ibid.
64. Rita Freedman and Nancy Dubler in ibid.
65. Myron Spector in ibid.
66. "Panel Issues Breast Implant Recommendations," FDA Talk Paper, November 15, 1991.
67. Robert Levier quoted in Philip J. Hilts, "FDA Panel Sees Need to Keep Breast Implants," *New York Times*, November 15, 1991.
68. "Plastic Surgeons Applaud FDA Panel's Call for Continued Access to Breast Implants," ASPRS news release, November 14, 1991.
69. "Plastic Surgeons Group Acts to Assure Breast Implant Patients of Information on Safety; Endorses Informed Consent Legislation, Explores Patient Registry," ASPRS news release, November 25, 1991.
70. Letter from Dow Corning CEO Daniel Hayes to plastic surgeons, November 27, 1991.
71. *Mariann Hopkins v. Dow Corning Corporation*, United States Court of Appeals for the Ninth Circuit, 33 F3d 1116 (1994); see also Byrne, *Informed Consent*.
72. Philip J. Hilts, "Implant Maker Accused on Data," *New York Times*, December 21, 1991.
73. According to John Byrne, one of Hopkins's lawyers leaked the documents to a reporter, who sent them on to Anderson. See Byrne, *Informed Consent*, 175.
74. The letter is described and portions are quoted verbatim in a letter from Dow Corning CEO Daniel Hayes to David Kessler, December 27, 1991.
75. Norman Anderson quoted in Hilts, "Implant Maker Accused."
76. Letter from Hayes to Kessler, December 27, 1991.
77. Hilts, "Implant Maker Accused."
78. Ibid.
79. Patti Scher and Marion Koch, *The Untold Truth about Breast Implants* (Charlotte, N.C.: 1993).
80. Quoted in *Mariann Hopkins v. Dow Corning Corporation*.
81. Philip J. Hilts, "FDA Seeks Halt in Breast Implants Made from Silicone," *New York Times*, January 7, 1992.
82. "European Surgeons Call for Independence from US Food and Drug Administration," *Journal of the National Cancer Institute* 85 (March 3, 1993): 353–354.
83. "FDA Calls for Moratorium on Silicone-Gel Breast Implants," "FDA, ASPRS Advise Women Not to Panic," *Breast Implant Update*, ASPRS newsletter, January 15, 1992.
84. Norman Cole quoted in Hilts, "FDA Seeks Halt."
85. Philip J. Hilts, "Makers of Silicone Breast Implants Say Data Show Them to Be Safe," *New York Times*, January 14, 1992.
86. "Independent Laboratory Report Shows No Implant-Related Adverse Reaction in Dog Study," *Dow Corning Corporate News* (newsletter), January 16, 1992.
87. Mitchell Karlan quoted in Hilts, "FDA Seeks Halt."
88. Ted Weiss quoted in ibid.

89. Hilts, "Makers . . . Say Data Show Them Safe."
90. "Defiance by Several Surgeons Is Reported," *New York Times*, January 13, 1992.
91. Robert Rylee quoted in Philip J. Hilts, "FDA Tells Company to Release Implant Data," *New York Times*, January 21, 1992.
92. "Company Will Release to the Public Additional Scientific Evidence Identified by the Agency; FDA Already in Possession of Studies; Dow Corning Renews Call for Expert Review," Dow Corning press release, January 21, 1992.
93. Ibid.
94. Matthew L. Wald, "An Ex-chemist's Formula for Dow Corning," *New York Times*, February 17, 1992.
95. ASPRS, "An Evaluation by Plastic Surgery of the 90 Documents Alleged by Commissioner David Kessler to Justify Moratorium on the Use of Silicone Gel Implants," prepared by members of the Silicone Implant Research Committee, Plastic Surgery Educational Foundation, February 1992 (unpublished document).
96. Felicity Barringer, "Action on Leader of FDA Urged," *New York Times*, February 27, 1992.
97. Tamar Lewin, "As Silicone Issue Grows, Women Take Agony and Anger to Court," *New York Times*, January 19, 1992.
98. Vedder, Price, Kaufman, and Kammholz, "Summary of Significant Decisions."
99. David Kessler in transcript of the FDA's General and Plastic Surgery Devices Panel Meeting, February 18–20, 1992. Hereafter cited as February Meeting.
100. Keith McKennon in ibid.
101. Robert Rylee in ibid.
102. James Baker in ibid.
103. Norman Cole in ibid.
104. Howard Tobin in ibid.
105. Sybil Niden Goldrich in ibid.
106. James Benson (FDA staffer) in ibid.
107. "FDA Recommends Limited Use of Silicone Gel-filled Breast Implants," *Medical Devices Bulletin*, FDA Center for Devices and Radiological Health, February 1992.
108. James Benson (FDA staffer) in February Meeting.
109. Vivian Snyder in ibid.
110. Nancy Dubler in ibid.
111. Mary McGrath in ibid.
112. Rita Freedman in ibid.
113. Sheryl Ruzek in ibid.
114. Marc Lappe in ibid.
115. Sidney Wolfe quoted in "Experts Suggest U.S. Sharply Limit Breast Implants," *New York Times*, February 21, 1992.
116. "Plastic Surgeons Disappointed in FDA Panel's Recommendation to Restrict Implant Access; Claim Decision Contradicts Hearing Discussion," ASPRS news release, February 20, 1992.
117. McKennon quoted in Byrne, *Informed Consent*, 196.

CHAPTER 2 THE PARADOX OF THE NATURAL

1. Vincenz Czerny, "Plastischer Ersatz der Brustdrüse durch ein Lipom," *Zentralblatt für Chirurgie* 22 (1895): 72.
2. Vincenz Czerny was born in 1842 in Germany. He studied medicine in Vienna, where he worked as assistant to some of the great names in the history of medicine. In 1871, he was made professor of surgery and director of a clinic in Freiburg. From the late nineteenth to the early years of the twentieth century, he published many articles on surgery and surgical technique, with an emphasis on operations for cancer. Czerny died in 1916.
3. Paul of Aegina quoted in Antony F. Wallace, *The Progress of Plastic Surgery: An Introductory History* (Oxford: Willem A. Meeuws, 1982), 55.
4. D. J. Hauben, "The History of Mammaplasty," *Acta Chirurgiae Plasticae* 27 (1985): 71–72.
5. Will Durston quoted in Gordon Letterman and Maxine Schurter, "Will Durston's 'Mammaplasty,'" *Plastic and Reconstructive Surgery* 53, 1 (1974): 48.

6. Ibid.
7. Ibid.
8. Wallace, *Progress of Plastic Surgery*, 155.
9. Ibid., 156
10. Ibid.; see also J. P. Lalardie and R. Mouly, "History of Mammaplasty," *Aesthetic Plastic Surgery* 2 (1978): 168–169.
11. Michel Foucault, *The Birth of the Clinic: An Archaeology of Medical Perception*, translated by A. M. Sheridan Smith (London: Tavistock, 1973).
12. Eugène Briau, "La Morphologie du sein féminin dans la classe ouvrière (essai d'hygiène Esthétique)," *Paris Médical* 54 (1924): 494–501.
13. M. Peugniez, "Esthétique du sein," *Revue Internationale de Médecine et de Chirurgie* 35 (1924): 128–130.
14. L. Dartiques, "Traitement chirurgical du prolapsus mammaire," *Archives Franco-Belges de Chirurgie* 28 (1925): 128. Translation is my own.
15. Frederick Strange Kolle, *Plastic and Cosmetic Surgery* (New York: D. Appleton, 1911), 209.
16. R. Gersuny, "Harte und weiche Paraffinprosthesen," *Zentralblatt für Chirurgie* 1 (1903): 1–5.
17. Max Thorek, *Plastic Surgery of the Breast and Abdominal Wall* (Springfield: Charles C. Thomas, 1942); Lalardie and Mouly, "History of Mammaplasty," 167–176.
18. Deborah N. Schalk, "The History of Augmentation Mammaplasty," *Plastic Surgical Nursing* 8, 3 (fall 1988): 88–90. The author cites sources that report the use of vegetable oil, lanolin, and beeswax.
19. Kolle, *Plastic and Reconstructive Surgery*, 215.
20. Charles C. Miller, "The Limitations and the Use of Paraffin in Cosmetic Surgery," *Wisconsin Medical Recorder* 11, 9 (1908): 277.
21. Davis quoted in H. Lyons Hunt, *Plastic Surgery of the Head, Face, and Neck* (Philadelphia: Lea and Febiger, 1926), 52.
22. Ibid., 49–50.
23. Weidman and Jeffries quoted in Hunt, *Plastic Surgery*, 54–55.
24. Karl-Heinrich Krohn, "Über Paraffinome der Mamma," *Zentralblatt für Chirurgie* 45 (1930): 2772–2781.
25. "Silicones May Aid Future Plastic Surgery," *Journal of the American Medical Association* 188, 9 (1964): 35.
26. Blair O. Rogers, "The Development of Aesthetic Plastic Surgery: A History," *Aesthetic Plastic Surgery* 1 (1976): 11.
27. Ibid.
28. Ibid; see also L. Dartiques, "De la greffe autoplastique libre aréolo-mamelonnaire combinée à la mammectomie bilatérale totale les raisons de sa prise," *Paris Chirurgie* 21 (1929): 11–19; Madame Noel and M. Lopez-Martinez, "Nouveaux Procédés chirurgicaux de correction du prolapsus mammaire." *Archives Franco-Belges de Chirurgie* 31 (1928): 138–153.
29. Schwarzmann, "Mammaplastik," *Der Chirurg* 2 (1930): 941.
30. See, for example, Briau, "La Morphologie"; Otto Frisch, "Die Indikationsstellung zur Korrektur von Formfehlern der weiblichen Brust," *Wiener Klinische Wochenschrift* 41 (1928): 640–641; Raymond Passot, "Atrophie mammaire," *La Presse Médicale* 37 (May 7, 1930): 627–628.
31. Passot, "Atrophie mammaire."
32. Noel, "Rapport des opérations esthétiques des seins avec les glandes ovariennes et mammaires," *La Bulletin Médical* (October 6, 1934): 611–613.
33. Lilian K. P. Farrar, "Nonmalignant Nodules of the Breast the Result of Prolapse," *Journal of the American Medical Association* 95, 18 (1930): 1329–1331.
34. Lois W. Banner, *American Beauty* (New York: Alfred A. Knopf, 1983).
35. Under a Galenic, or humoral, model of disease, breast cancer was conceptualized as a pathological accumulation, or congestion, of black bile, one of the four humors. In this schema, breast cancer was essentially a systemic disease with a local manifestation. The treatments aimed to relieve the local congestion by drawing out the bile—perhaps by cauterizing the lesion or by applying a poultice—and to treat the whole

system—perhaps with diet or herbs. By the beginning of the nineteenth century, the medical conceptualization of disease had become one of solidism, or a concern with tissues and their structure. Breast cancer was then believed to be a disease of disordered tissue, a local condition best treated by excising the lesion. For more about changing conceptualizations of breast cancer, see Daniel de Moulin, *A Short History of Breast Cancer* (Boston: Martinus Nijhoff, 1983).

36. William Halsted quoted in Joan Austoker, "The 'Treatment of Choice': Breast Cancer Surgery, 1860–1985," *Bulletin of the Society for the Social History of Medicine* 37 (1985): 101.
37. Ibid; see also Theresa Montini and Sheryl Ruzek, "Overturning Orthodoxy: The Emergence of Breast Cancer Treatment Policy," *Research in the Sociology of Health Care* 8 (1989): 3–32.
38. J. Moreau, "Plastique après amputation du sein pour cancer," *Journal Médical de Bruxelles* 3 (January 1908): 33–37.
39. Before cell theory became the dominant medical model, cases of breast cancer were not diagnosed until women presented with a visible lesion, either a deformation of the skin from a tumor, or an open sore. At this point, the cancer was far advanced and treatment (of any kind) rarely was effective. With the cellular theory of breast cancer, however, came the realization that what began as a single cell could be recognized and treated earlier. Along with the Halsted radical came efforts to educate physicians and the public about early detection and early treatment. (See, e.g., Joseph Colt Bloodgood, "What Every Doctor Should Know About the Breast," *Archives of Clinical Cancer Research* 1, 4 [1925]: 1–29.) This shift had several implications: Survival times after surgery increased. More women underwent mastectomies. Younger women became candidates for the surgery.
40. Moreau, "Plastique après amputation"; Bahman Teimourian and Mehdi N. Adham, "Louis Ombrédanne and the Origin of Muscle Flap Use for Immediate Breast Mound Reconstruction," *Plastic and Reconstructive Surgery* 72, 6 (1983): 905–910.
41. H. Morestin, "De l'autoplastie par déplacement du sein," *Archives Générale de Médecine* 2 (1903): 2689–2698.
42. L. Ombrédanne, "Restauration autoplastique du sein," *La Tribune Médicale* 38 (1906): 325–326.
43. Teimourian and Adham, "Louis Ombrédanne and the Origin of Muscle Flap Use."
44. Frisch, "Die Indikationsstellung," 640. Translation by Dolly Jacobson.
45. Passot, "Atrophie mammaire." Translation is my own.
46. H. O. Bames, "Plastic Reconstructive of the Anomalous Breast," *Revue de Chirurgie Structive* 55 (July 1936): 293.
47. Ibid., 294.
48. Thorek, *Plastic Surgery of the Breast and Abdominal Wall*, 3.
49. Morton I. Berson, "Derma-Fat-Fascia Transplants Used in Building Up of the Breast," *Surgery* 15 (March 1944): 451–456.
50. Ibid., 456.
51. Before 1950, crude rates of mortality due to breast cancer grouped by decade show a steady but slow increase. (U.S. Department of Health, Education, and Welfare, National Center for Health Statistics, *Vital Statistics Rates in the United States, 1940–1960*, Public Health Service Publication #1677 [Washington, D.C.: U.S. Government Printing Office, 1968].) No national figures on incidence are available. Between 1950, when data were first collected in this form, and 1980, compilations of national data on breast cancer incidence and mortality show a trend of increasing incidence—reflecting, most likely, increased disease surveillance—and steady age-adjusted mortality rates. (U.S. Department of Health and Human Services, *Atlas of United States Cancer Mortality*, Publication #90–1582 [Washington, D.C.: U.S. Government Printing Office, 1987].) Interpretation of the two trends indicates that although more women were being diagnosed, no more were dying each year, suggesting that simply because women were being diagnosed earlier, they were appearing to live longer—a phenomenon known as lead time bias.
52. Harold I. Harris, "Automammaplasty," *Journal of the International College of Surgeons* 12, 6 (1949): 827–839.
53. The "accidental" mastectomy for benign conditions was facilitated by the "one step" surgical protocol developed alongside the Halsted radical. In that protocol, the woman

was anesthetized, an incision made, and a biopsy of the suspected tissue sent to the hospital histology lab for a frozen section. Within a half hour, the pathologist could read the slide. A determination of malignancy meant that the woman would undergo a radical mastectomy. It was all too common, however—as Harris described—for a more thorough pathological examination of the resected tissue to contradict the original reading of the frozen section. The "one step" procedure—in which the woman went under anesthesia not knowing "if she would wake up with one breast or two"—survived well into the early 1980s. Changing this protocol was one of the aims of early breast cancer activists. See Rose Kushner, *Breast Cancer: A Personal History and an Investigative Report* (New York: Harcourt Brace Jovanovich, 1975).

54. Willard Bartlett, "An Anatomic Substitute for the Female Breasts," *Annals of Surgery* (Philadelphia) 66 (1917): 208–211.
55. Ibid., 208.
56. Carl O. Rice and J. H. Strickler, "Adenomammectomy for Benign Breast Lesions," *Surgery, Gynecology, and Obstetrics* 93 (1951): 759–762.
57. R. R. McGregor, "Commentary," *Bulletin of the Dow Corning Center for Aid to Medical Research* 2, 1 (1960).
58. Germain Gillet, "Traitement chirurgical de l'insuffisance plastique mammaire," *La Presse Médicale* 62, 63 (1954): 1318–1319.
59. Pierre Lombard and Jacques Frailong, "Le Sein sur plexiglass," *Bulletin de l'Acadèmic Nationale de Médecine* 137 (June 16, 1953): 327–328 .
60. John H. Grindlay and John M. Waugh, "Plastic Sponge Which Acts as a Framework for Living Tissue," *A.M.A. Archives of Surgery* 63 (1951): 288–297.
61. Ibid., 288.
62. Ibid., 295.
63. Ibid., 296.
64. B. S. Oppenheimer, Enid T. Oppenheimer, and Arthur Purdy Stout, "Sarcomas Induced in Rodents by Imbedding Various Plastic Films," *Proceedings of the Society for Experimental Biology and Medicine* 79 (1952): 366–369.
65. This clinical use was concurrent with the experimental laboratory studies. Although the first clinical reports did not appear until about 1953, the authors claimed to have been using the materials since 1950.
66. Pangman held at least three patents for variants of the Ivalon implant. (Patents 3,559,214; 3,189,921; and 2,842,775 in the *Official Gazette*.)
67. W. John Pangman and Robert M. Wallace, "The Use of Plastic Prostheses in Breast Plastic and Other Soft Tissue Surgery," *Western Journal of Surgery, Obstetrics and Gynecology* 63 (August 1955): 508.
68. An audiotape of this conversation, which took place on October 28, 1979, is stored at the National Archives of Plastic Surgery.
69. Pangman and Wallace, "Use of Plastic Prostheses," 512.
70. Garry S. Brody, "Overview of Augmentation Mammaplasty," in *Plastic and Reconstructive Surgery of the Breast*, edited by R. Barrett Noone (Philadelphia: B. C. Decker, 1991), 120.
71. Robert Alan Franklyn, "Die Verwendung von Silikonschwamm-Prosthesen in kosmetischer Chirurgie," *Zentralblatt für Chirurgie* 80 (1955): 1553–1559. Translation by Dolly Jacobson.
72. Cited in C. W. Trowbridge, "The Use of Plastics in Reconstructing the Female Breast," *Journal of the Medical Association of Alabama* 29, 9 (1960): 331.
73. Franklyn quoted in "The Business of Bolstering Bosoms," *JAMA* 153, 13 (1953): 1200–1201.
74. Ibid., 1200.
75. Ibid., 1201.
76. Ibid., 1200.
77. Simon Fredricks, "Management of Mammary Hypoplasia," in *Plastic and Reconstructive Surgery of the Breast*, edited by Robert M. Goldwyn (Boston: Little, Brown, 1976), 388.
78. Harold I. Harris, "Correction of the Hypoplastic Breast by the Use of Dermo-Fat and Plastic Ivalon Sponge," and Julio Utermin Aguirre, "Polyethylene in Hypotrophic Breasts," in *Transactions of the First International Congress of Plastic Surgery* (Stockholm and Uppsala, 1955).

79. Robert M. Goldwyn, "Comment," in *Long Term Results in Plastic and Reconstructive Surgery*, vol. 2, edited by Robert M. Goldwyn (Boston: Little, Brown, 1980).
80. James Barrett Brown, Minot Fryer, and Milton Lu, "Polyvinyl and Silicone Compounds as Subcutaneous Prostheses: Laboratory and Clinical Investigations," *A.M.A. Archives of Surgery* 66 (1954): 746.
81. Gerald Brown O'Connor, Mar Watson McGregor, and Allen H. Long, "Mastoplasty in the Last Decade: A Review," *Plastic and Reconstructive Surgery* 17, 6 (1956): 487.
82. William Kiskadden, "Operations on Bosoms Dangerous," *Plastic and Reconstructive Surgery* 15, 1 (1955): 79.
83. Ibid., 80.
84. W. C. Hueper, "Carcinogenic Studies on Water-Soluble and Insoluble Macromolecules," *A.M.A. Archives of Pathology* 67 (June 1959): 13–39.
85. Cuthbert E. Dukes and Bernard C. V. Mitchley, "Polyvinyl Sponge Implants: Experimental and Clinical Observations," *British Journal of Plastic Surgery* 21 (1962): 235.
86. Then, as now, licensed physicians needed no specific certification in plastic surgery to perform cosmetic or reconstructive procedures. Given the contemporaneous hostility of the plastic surgery establishment to artificial materials, doctors who were not members of the plastic surgeons' professional society may have been more likely to have been using synthetic implants.
87. Harold I. Harris, "Survey of Breast Implants from the Point of View of Carcinogenesis," *Plastic and Reconstructive Surgery* 28, 1 (1961): 81–83. Harris fails to report how many surveys were sent out, but out of 294 responses, 185 surgeons reported placing a total of 16,660 implants. Although most respondents indicated that they had performed between 1 and 10 implantations, one surgeon had performed 3,000 procedures.
88. Reuven K. Snyderman and Jesus G. Lizardo, "Statistical Study of Malignancies Found before, during, or after Routine Breast Plastic Operations," *Plastic and Reconstructive Surgery* 25, 3 (1960): 253–256.
89. This list, developed by Scales and cited most often, is described in Marvin S. Arons, Seymour M. Sabesin, and Robert R. Smith, "Experimental Studies with Etheron Sponge," *Plastic and Reconstructive Surgery* 28, 1 (1961): 72.
90. W. G. Holdsworth, "A Method of Reconstructing the Breast," *British Journal of Plastic Surgery* 9 (1957): 161.
91. Francis A. Marzoni, Samuel E. Upchurch, and C. J. Lambert, "An Experimental Study of Silicone as a Soft Tissue Substitute," *Plastic and Reconstructive Surgery* 24, 6 (1959): 608.
92. Vaughn Dermergian, "Experiences with the Newer Subcutaneous Implant Materials," *Surgical Clinics of North America* 43, 5 (1963): 1313.
93. Fredricks, "Management of Mammary Hypoplasia," 367.
94. Banner, *American Beauty*.
95. O'Connor, McGregor, and Long, "Mastoplasty," 484.

CHAPTER 3 A PLEASING ENLARGEMENT FROM WITHIN

1. James Barrett Brown, Minot P. Fryer, and David A. Ohlwiler, "Study and Use of Synthetic Materials such as Silicones and Teflon as Subcutaneous Prostheses," *Plastic and Reconstructive Surgery* 26, 3 (1960): 376.
2. Francis Marzoni, Samuel E. Upchurch, and C. J. Lambert, "An Experimental Study of Silicone as a Soft Tissue Substitute," *Plastic and Reconstructive Surgery* 24, 6 (1959): 600–608.
3. Barrett Brown, Fryer, and Ohlwiler, "Silicones and Teflon as Subcutaneous Prostheses."
4. Benjamin F. Edwards, "Teflon-Silicone Breast Implants," *Plastic and Reconstructive Surgery* 32, 5 (1963): 519–526.
5. James Barrett Brown, Minot P. Fryer, Peter Randall, and Milton Lu, "Silicones in Plastic Surgery," *Plastic and Reconstructive Surgery*, 12 (1953): 374–376; R. De Barondes, William D. Judge, Charles G. Towne, and Metta L. Baxter, "The Silicones in Medicine," *Military Surgeon* 106 (1950): 379–387.
6. Howard W. Post, *Silicones and Other Organic Silicon Compounds* (New York: Reinhold, 1949).

7. The history that follows comes from Earl L. Warrick, *Forty Years of Firsts: The Recollections of a Dow Corning Pioneer* (New York: McGraw-Hill, 1990).
8. Ibid., 87.
9. De Barondes et al., "Silicones in Medicine," 384.
10. Silas Braley, "Use of Silicones in Plastic Surgery," *Archives of Otolaryngology* 78 (November 1963): 669–675. Dow Corning produced silicones with a variety of purity levels, ranging from industrial grade to medical grade.
11. T. D. Cronin and F. J. Gerow, "Augmentation Mammaplasty: A New 'Natural Feel' Prosthesis," in *Transactions of the Third International Congress of Plastic Surgery* (Amsterdam: Excerpta Medica Foundation, 1964), 41.
12. Simon Fredricks, "Management of Mammary Hypoplasia," in *Plastic and Reconstructive Surgery of the Breast*, edited by Robert M. Goldwyn (Boston: Little, Brown, 1976), 388.
13. Cronin and Gerow, "'Natural Feel' Prosthesis," 42.
14. Ibid., 43.
15. Ibid., 47.
16. Ibid.
17. "Silicones May Aid Future Plastic Surgery," *JAMA* 188, 9 (1964): 35.
18. Ralph Blocksma and Silas Braley, "The Silicones in Plastic Surgery," *Plastic and Reconstructive Surgery* 35, 4 (1965): 366.
19. Ibid., 369.
20. H. Koehlin, "Prosthèses mammaire," *Praxis* 54 (1965): 894–895; John B. Erich, "Augmentation Mammaplasty," *Mayo Clinic Proceedings* 40 (May 1965): 397–402.
21. John R. Lewis, "The Augmentation Mammaplasty," *Plastic and Reconstructive Surgery* 35, 1 (1965): 51–59.
22. Joseph R. Zbylski and Robert W. Parsons, "A Method of Adenectomy and Breast Reconstruction for Benign Disease," *Plastic and Reconstructive Surgery* 37, 1 (1966): 38–41; Alex P. Kelly et al., "Complications of Subcutaneous Mastectomy and Replacement by the Cronin Silastic Mammary Prosthesis," *Plastic and Reconstructive Surgery* 37, 5 (1966): 438–445.
23. Erich, "Augmentation Mammaplasty," 400.
24. Lewis, "Augmentation Mammaplasty."
25. When, in the context of the controversy, the history of silicone use came to be of interest, several observers implied that implants were a response to silicone injection, that the technology represented a modification of a technique found to be dangerous. (See, e.g., the testimony of Norman Anderson in *Weiss Hearing* [full citation in chapter 1, note 3] and John A. Byrne, *Informed Consent* [New York: McGraw Hill, 1996].) In fact, in the United States at least, the two procedures were going on at the same time.
26. Norman Anderson recounts this history in *Weiss Hearing*. See also Judy Foreman, "Women and Silicone: A History of Risk," *Boston Globe*, January 19, 1992.
27. H. D. Kagan, "Sakurai Injectable Silicone Formula," *Archives of Otolaryngology* 78 (1963): 663–667.
28. Byrne, *Informed Consent*, 44.
29. *Promotion of Drugs and Medical Devices for Unapproved Uses, Hearing before the Human Resources and Intergovernmental Relations Subcommittee of the Committee on Government Operations*, House of Representatives, 102nd Cong., 1st sess., June 11, 1991. Before the FDA was able to have silicone classified as a drug, however, plastic surgeons were studying ways to use the material for breast augmentation. See, for example, Herbert Conway and Dicran Goulian, "Experience with an Injectable Silastic RTV as a Subcutaneous Prosthetic Material," *Plastic and Reconstructive Surgery* 32, 3 (1963): 294–302.
30. Franklin L. Ashley et al., "The Present Status of Silicone Fluid in Soft Tissue Augmentation," *Plastic and Reconstructive Surgery* 39, 4 (1967): 411–420.
31. Nahum Ben-Hur et al., "Local and Systemic Effects of Dimethylpolysiloxane Fluid in Mice," *Plastic and Reconstructive Surgery* 39, 4 (1967): 423–426.
32. Thomas D. Rees, Donald L. Ballantyne, and Gail A. Hawthorne, "Silicone Fluid Research: A Follow-Up Summary," *Plastic and Reconstructive Surgery* 46, 1 (1970): 50–56.

33. Ashley, "Present Status," 415.
34. Ibid., 411.
35. See Byrne, *Informed Consent*, 42–44.
36. Testimony of Norman Anderson in *Weiss Hearing*, 31.
37. "The AMA Has Issued a Warning to Women," *JAMA* 217, 11 (1971): 1452.
38. Al Reinart, "Doctor Jack Makes His Rounds," *Esquire*, May 1975, 114, 116, 160–163.
39. See chapter 5 for a detailed discussion of these warnings. Note also that many of them were about injections with materials other than silicone.
40. Louis H. Winer et al., "Tissue Reactions to Injected Silicone Liquids," *Archives of Dermatology* 90 (December 1964): 588–593; W. St. C. Symmers, "Silicone Mastitis in 'Topless' Waitresses and Some Other Varieties of Foreign-Body Mastitis," *British Medical Journal* 3 (July 1968): 19–22; Jerome S. Nosanchuk, "Silicone Granuloma in Breast," *Archives of Surgery* 97 (October 1968): 583–585; C. Hal Chaplin, "Loss of Both Breasts from Injection of Silicone (with Additive)," *Plastic and Reconstructive Surgery* 44, 5 (1969): 447–450; Khoo Boo-Chai, "The Complications of Augmentation Mammaplasty by Silicone Injection," *British Journal of Plastic Surgery* 22, 3 (1969): 281–285; Richard Ellenbogen, Rita Ellenbogen, and Leonard Rubin, "Injectable Fluid Silicone Therapy: Human Morbidity and Mortality," *JAMA* 234, 3 (1975): 308–309.
41. One author noted a whole range of substances inserted into the breast: "paraffin waxes, beeswax, silicone wax, silicone fluid, shellac, shredded oiled-silk fabric, silk tangle, glazer's putty, spun glass, and epoxy resin." Symmers, "Silicone Mastitis," 19.
42. Ibid.
43. Nosanchuk, "Silicone Granuloma in the Breast," 585.
44. Reinart, "Doctor Jack," 160.
45. Joseph E. Murray, "Editorial: Factors for Safety in Use of Silicone," *Plastic and Reconstructive Surgery* 39, 4 (1967): 427.
46. See *Promotion of Drugs and Medical Devices for Unapproved Uses*.
47. Letter from Carl A. Larson (FDA Office of Device Evaluation) to Arthur Rathjen (Dow Corning), September 21, 1990. Reproduced in *Promotion of Drug and Medical Devices for Unapproved Uses*, 207–208.
48. See *Promotion of Drugs and Medical Devices for Unapproved Uses*; one such doctor was Norman Orentreich, a prominent New York City dermatologist. Nancy Reagan was, reputedly, one of Orentreich's patients. In the early 1990s, there was speculation that FDA investigations of Orentreich had been quashed due to political pressure.
49. Fritz Bischoff, "Organic Polymer Biocompatility and Toxicology," *Clinical Chemistry* 18, 9 (1972): 888.
50. Ibid.
51. "The Beauty, the Beast, and the Breast," *JAMA* 223, 6 (1973): 683.
52. For example, Michael Gurdin and Gene A. Carlin, "Complications of Breast Implantations," *Plastic and Reconstructive Surgery* 40, 6 (1967): 530–533; Thomas D. Cronin and Roger L. Greenberg, "Our Experiences with the Silastic Gel Breast Prosthesis," *Plastic and Reconstructive Surgery* 46, 1 (1970): 1–7; John E. Williams, "Experiences with a Large Series of Silastic Breast Implants," *Plastic and Reconstructive Surgery* 49, 3 (1972): 253–262.
53. Paule Regnault, "Indications for Breast Augmentation," *Plastic and Reconstructive Surgery* 40, 6 (1967): 524–529.
54. Cronin and Greenberg, "Silastic Gel Breast Prosthesis."
55. Williams, "Experiences with a Large Series," 254.
56. Joseph R. Zbylski and Robert W. Parsons, "A Method of Adenectomy and Breast Reconstruction for Benign Disease," *Plastic and Reconstructive Surgery* 37, 1 (1966): 38.
57. Ibid., 41.
58. Alex P. Kelly, Herbert S. Jacobson, Joseph L. Fox, and Henry Jenny, "Complications of Subcutaneous Mastectomy and Replacement by the Cronin Silastic Mammary Prothesis," *Plastic and Reconstructive Surgery* 37, 5 (1966): 438.
59. J. P. Lalardie and D. Morel-Fatio, "Mammectomie totale sous-cutanee suivie de reconstruction immediate ou secondaire," *Chirurgie* 98 (1970): 651. Translation is my own.

60. Thomas H. Bill Allen, "Subcutaneous Mammectomy: New Hope for Benign Breast Disease," *Journal of the Arkansas Medical Society* 69, 5 (1972): 153.
61. Reuven K. Snyderman and Randolph Guthrie, "Reconstruction of the Female Breast Following Radical Mastectomy," *Plastic and Reconstructive Surgery* 47, 6 (1971): 565.
62. John N. Simons, "Nonaesthetic Augmentation Mammaplasty," and Charles Horton, Jerome Adamson, and Richard Mladick, "Breast Reconstruction after Surgery and Trauma," in *Symposium on Aesthetic Surgery of the Face, Eyelid, and Breast*, edited by Frank W. Masters and John R. Lewis (St. Louis: C. V. Mosby, 1972).
63. Williams, "Experiences with a Large Series."
64. Gurdin and Carlin, "Complications."
65. Ibid., 532–533.
66. Bromley S. Freeman, "Successful Treatment of Some Fibrous Envelope Contractures Around Breast Implants," *Plastic and Reconstructive Surgery* 50, 2 (1972): 107–113.
67. Blocksma and Braley, "Silicones in Plastic Surgery."
68. Williams, "Experiences with a Large Series."
69. Cronin and Greenberg, "Silastic Gel Breast Prosthesis."
70. Thomas M. Biggs, Jean Cukier, and L. Fabian Worthing, "Augmentation Mammaplasty: A Review of 18 Years," *Plastic and Reconstructive Surgery* 69, 3 (1982): 445–450.
71. H. G. Arion, "Retromammary Prostheses," *Comptes Rendus de la Société Francaises de Gynécologie* 5 (May 1965).
72. Kuros Tabari, "Augmentation Mammaplasty with Simaplast Implant," *Plastic and Reconstructive Surgery* 44, 5 (1969): 468–470.
73. Cronin and Greenberg, "Silastic Gel Breast Prosthesis."
74. Franklin L. Ashley, "A New Type of Breast Prosthesis: Preliminary Report," *Plastic and Reconstructive Surgery* 45, 5 (1970): 421–424.
75. W. C. Dempsey and W. D. Latham, "Subpectoral Implants in Augmentation Mammaplasty: Preliminary Report," *Plastic and Reconstructive Surgery* 42, 5 (1968): 515–521.
76. James Baker quoted in "Breast Building: A Boon with Hidden Hazards," *Medical World News*, 14 (1973): 60.
77. Fredricks, "Mammary Hypoplasia."
78. *Augmentation Mammaplasty as an Office Procedure*. I viewed the film at the National Archives of Plastic Surgery.
79. Byrne, *Informed Consent*, 61.
80. Ibid., 66.
81. John E. Hoopes, Milton T. Edgerton, and William Shelley, "Organic Synthetics for Augmentation Mammaplasty: Their Relation to Breast Cancer," *Plastic and Reconstructive Surgery* 39, 3 (1967): 263–270; Tibor de Cholnoky, "Augmentation Mammaplasty: Survey of Complications in 10,941 Patients by 265 Surgeons," *Plastic and Reconstructive Surgery* 45, 6 (1970): 573–577.
82. de Cholnoky, "Survey of Complications," 577.
83. Robby Meijer, "When Your Patient Asks about Plastic Surgery," *Medical Times* 99, 9 (1971): 117.
84. Gurdin and Carlin, "Complications," 159.
85. Silas Braley, "The Present Status of Mammary Implant Materials," in *Symposium on Aesthetic Surgery of the Face, Eyelid, and Breast*, 149. Note, however, that this does not necessarily mean that fifty thousand different women had at least one breast implant.
86. Erich, "Augmentation Mammaplasty," 397.
87. Braley, "Present Status," 150.
88. Ibid., 151.
89. Nicholas Regush, "Toxic Breasts," *Mother Jones*, January–February 1992, 25–31.
90. Thomas D. Rees, Carl L. Guy, and Richard J. Coburn, "The Use of Inflatable Breast Implants," *Plastic and Reconstructive Surgery* 53, 2 (1974): 609–615.
91. Carl H. Dahl, M. Edward Baccari, and Parvis Arfai, "A Silicone Gel Inflatable Mammary Prosthesis," *Plastic and Reconstructive Surgery* 53, 2 (1974): 234–235.

92. John. H. Hartley, "Specific Applications of the Double Lumen Prosthesis," *Clinics in Plastic Surgery* 3, 2 (1976): 247–263.
93. A. Richard Grossman, "The Current Status of Augmentation Mammaplasty," *Plastic and Reconstructive Surgery* 52, 1 (1973): 1–7.
94. Cited by Nir Kossovsky during the February 1992 FDA advisory panel meeting.
95. Eugene H. Courtiss, Richard C. Webster, and Malvin F. White, "Selection of Alternatives in Augmentation Mammaplasty," *Plastic and Reconstructive Surgery* 54, 5 (1974): 552–557.
96. Ibid., 552.
97. Jack Penn, "A Perspective and Technique for Breast Reduction," in *Plastic and Reconstructive Surgery of the Breast*, edited by Robert M. Goldwyn (Boston: Little, Brown, 1976), 184.
98. James Baker, "Augmentation Mammaplasty," in *Symposium on Aesthetic Surgery of the Breast*, edited by John Q. Owsley and Rex A. Peterson (St. Louis: C. V. Mosby, 1978), 256.
99. Ibid., 264.
100. Fredricks, "Management of Mammary Hypoplasia," 395.
101. John T. Hueston, "Unilateral Agenesis and Hypoplasia: Difficulties and Suggestions," in *Plastic and Reconstructive Surgery of the Breast*, 361.
102. Bernard L. Kaye, "Micromastia v. Macromastia," *Medical Aspects of Human Sexuality* 7 (August 1973): 353.
103. Theresa Montini and Sheryl Ruzek, "Overturning Orthodoxy: The Emergence of Breast Cancer Treatment Policy," *Research in the Sociology of Health Care* 8 (1989): 3–32.
104. See the National Center for Health Statistics report series *Surgical Operations in Short Stay Hospitals* for 1973, 1975, and 1978 (Washington, D.C.: U.S. Government Printing Office); National Center for Health Statistics, *Use of Health Services for Disorders of the Female Reproductive System. United States, 1977–78* (Washington, D.C.: U.S. Government Printing Office: 1982); and National Center for Health Statistics, *Surgical and Nonsurgical Procedures in Short-Stay Hospitals, 1979* (Washington, D.C.: U.S. Government Printing Office, 1979).
105. Sharon Batt, *Patient No More: The Politics of Breast Cancer* (Prince Edward Island, Canada: gynergy, 1994).
106. Thomas D. Cronin, Joseph Upton, and James M. McDonough, "Reconstruction of the Breast after Mastectomy," *Plastic and Reconstructive Surgery* 59, 1 (1977): 1–14.
107. Kuros Tabari, "Breast Reconstruction after Radical Mastectomy—an Approach to Rehabilitative Surgery," *Annals of Plastic Surgery* 7, 3 (1981): 222–227.
108. John E. Woods, "Breast Reconstruction after Mastectomy," *Surgery, Gynecology, and Obstetrics* 150 (June 1980): 869–874. For descriptions of the most elaborate of these procedures, see Robert M. Goldwyn, "Breast Reconstruction after Mastectomy," *New England Journal of Medicine* 317, 27 (1987): 1711–1714.
109. Chedomir Radovan, "Breast Reconstruction after Mastectomy Using the Temporary Expander," *Plastic and Reconstructive Surgery* 69, 2 (1982): 195–206.
110. John Bostwick, "Breast Reconstructions after Mastectomy: Recent Advances," *Cancer* 66 (supplement) (September 15, 1990): 1402–1411.
111. Charles E. Horton, "Immediate Reconstruction following Mastectomy for Cancer," in *Symposium on Reconstruction of the Breast after Mastectomy*, edited by Leonard R. Rubin, *Clinics in Plastic Surgery* 6, 1 (1979).
112. John T. Hueston, "Immediate Reconstruction after Radical Mastectomy," in *Symposium on Aesthetic Surgery of the Breast.*
113. C. Patrick Maxwell, "Selection of Secondary Breast Reconstruction Procedures," in *Symposium on Advances in Breast Reconstruction*, edited by Michael Scheflan, *Clinics in Plastic Surgery* 11, 2 (1984): 253.
114. Michael Scheflan, "Foreword," in ibid., 229.
115. Ibid.
116. Goldwyn, "Breast Reconstruction."
117. Ibid.
118. Dennis M. Deapen et al., "The Relationship between Breast Cancer and Augmentation Mammaplasty: An Epidemiologic Study," *Plastic and Reconstructive Surgery* 77, 3 (1986): 361–367.
119. For example, Paul P. Pickering, John E. Williams, and Thomas R. Vecchione, "Aug-

mentation Mammaplasty," in *Long Term Results in Plastic and Reconstructive Surgery*, vol. 2, edited by Robert M. Goldwyn (Boston: Little, Brown, 1980); Biggs, Cukier, and Worthing, "Augmentation Mammaplasty," 445–450; Thomas Cronin, Myron Persoff, and Joseph Upton, "Augmentation Mammaplasty: Complications and Etiology," in *Symposium on Aesthetic Surgery of the Breast.*

120. Cronin, Persoff, and Upton, "Augmentation Mammaplasty," 272.
121. Fredricks, "Management of Mammary Hypoplasia."
122. Ross Rudolph et al., "Myofibroblasts and Free Silicon around Breast Implants," *Plastic and Reconstructive Surgery* 62, 2 (1978): 19.
123. Thomas M. Biggs and R. Scott Yarish, "Augmentation Mammaplasty: A Comparative Analysis," *Plastic and Reconstructive Surgery* 85, 3 (1990): 368.
124. Daniel Weiner, Adrien E. Aiache, and Lester Silver, "A New Soft, Round Silicone Gel Breast Implant," *Plastic and Reconstructive Surgery* 62, 2 (1978): 177.
125. Jaime Planas, "Mammary Augmentation—Surgical Techniques, Evaluation of Results, and Complications," in *Symposium on Breast Surgery, Clinics in Plastic Surgery* 3, 2 (1976): 242.
126. Rudolph et al., "Myofibroblasts."
127. Garry S. Brody, "Fact and Fiction about Breast Implant 'Bleed'" (editorial), *Plastic and Reconstructive Surgery* 60, 4 (1977): 615.
128. Ibid., 615–616.
129. J. Edson Price and Donald E. Parker, "Initial Experience with 'Low Bleed' Breast Implants," *Aesthetic Plastic Surgery* 7 (1983): 255–256.
130. Regush, "Toxic Breasts."
131. James L. Baker, Roger L. Bartels, and William M. Douglas, "Closed Compression Technique for Rupturing a Contracted Capsule around a Breast Implant," *Plastic and Reconstructive Surgery* 58, 2 (1976): 133.
132. Biggs and Yarish, "Comparative Analysis," 369.
133. Boyd Burkhardt, "Capsular Contracture: Hard Breasts, Soft Data," *Clinics in Plastic Surgery* 15, 4 (1988): 521–532.
134. Ibid., 529.
135. Pickering, Williams, and Vecchione, "Augmentation Mammaplasty," 698.
136. Garry S. Brody, "Overview of Augmentation Mammaplasty," in *Plastic and Reconstructive Surgery of the Breast*, edited by R. Barrett Noone (Philadelphia: B. C. Decker, 1991), 124.
137. ASPRS statistics for 1988 report 71,720 augmentations and 34,210 reconstructions. For 1990, the numbers are 89,402 augmentations and 42,888 reconstructions. Note that these are underestimates of the likely total number of surgeries performed since they reflect only those procedures performed by the members of the organization.
138. Ronald E. Everson, "National Survey Shows Overwhelming Satisfaction with Breast Implants" (letter to the editor), *Plastic and Reconstructive Surgery* 88, 3 (1991): 546–547.
139. Brody, "Overview," 124.

CHAPTER 4 THE MEDICAL CONSTRUCTION OF NEED (OR, THE "PSYCHOLOGY OF THE FLAT-CHESTED WOMAN")

1. Of course, like need, desire is also socially constructed. For examples of how scholars have explored the social construction of women's desire for cosmetic surgery, including breast implants, see: Carole Spitzak, "The Confession Mirror: Plastic Images for Surgery," *Canadian Journal of Political and Social Theory* 12, 1–2 (1988): 38–50; Diana Dull and Candace West, "Accounting for Cosmetic Surgery: The Accomplishment of Gender," *Social Problems* 38, 1 (1991): 54–70; Kathryn Pauly Morgan, "Women and the Knife: Cosmetic Surgery and the Colonization of Women's Bodies," *Hypatia* 6, 3 (1991): 25–53; Kathy Davis, *Reshaping the Female Body: The Dilemma of Cosmetic Surgery* (New York: Routledge, 1995).
2. I also explore this argument in "The Socially Constructed Breast: Breast Implants and the Medical Construction of Need," *American Journal of Public Health* 88, 8 (1998): 1254–1261.

3. Elizabeth Haiken traces this development in *Venus Envy: A History of Cosmetic Surgery* (Baltimore: Johns Hopkins University Press, 1997).
4. Madame Noel and M. Lopez-Martinez, "Nouveaux Procédés chirurgicaux de correction du prolapsus mammaire," *Archchives Franco-Belges de Chirurgie* 31 (1928): 138. Translation is my own.
5. Max Thorek, *Plastic Surgery of the Breast and Abdominal Wall* (Springfield: Charles C. Thomas, 1942), vii.
6. Ibid., 48.
7. Ibid., 50.
8. H. O. Bames, "Breast Malformations and a New Approach to the Problem of the Small Breast," *Plastic and Reconstructive Surgery* 5 (1950): 506.
9. Jacques Maliniac, *Breast Deformities and Their Repair* (Baltimore: Waverly Press, 1950), vii.
10. Thorek, *Plastic Surgery of the Breast*, 51.
11. H. O. Bames, "Plastic Reconstructive of the Anomalous Breast," *Revue du Chirurgie Structive* (July 1936): 293.
12. H. O. Bames, "Correction of Abnormally Large or Small Breasts," *Southwestern Medicine* 25 (January 1941): 10.
13. Bames, "The Anomalous Breast," 293.
14. Maliniac, *Breast Deformities*, 5.
15. J. Eastman Sheehan, "Foreword," in Thorek, *Plastic Surgery of the Breast*, ix.
16. Bames, "Abnormally Large or Small Breasts," 13.
17. Thorek, *Plastic Surgery of the Breast*, 51.
18. Maliniac, *Breast Deformities*, vii.
19. Robert Alan Franklyn quoted in "The Business of Bolstering Bosoms," *JAMA* 153, 13 (1953): 1200.
20. W. John Pangman and Robert M. Wallace, "Use of Plastic Prostheses in Breast Plastic and Other Soft Tissue Surgery," *Western Journal of Surgery, Obstetrics and Gynecology* 63 (August 1955): 503.
21. "Psychology of the Flat-Chested Woman" is the title of a chapter by John Hoopes and Norman Knorr in *Symposium on Aesthetic Surgery of the Face, Eyelid, and Breast*, edited by Frank W. Masters and John R. Lewis (St. Louis: C. V. Mosby, 1972).
22. Karl A. Menninger, "Somatic Correlations with the Unconscious Repudiation of Femininity in Women," *Bulletin of the Menninger Clinic* 3 (1939): 106–121.
23. Ibid., 111.
24. Ibid., 109.
25. Karl Stern, Denis Doyon, and Roger Racine, "Preoccupation with the Shape of the Breast as a Psychiatric Symptom in Women," *Canadian Psychiatric Association Journal* 4, 4 (1959): 252.
26. Alexander Grinstein, "Some Comments on Breast Envy in Women," *Journal of the Hillside Hospital* 11, 4 (1962): 172.
27. Ibid., 176.
28. Ibid., 172.
29. Ibid., 176.
30. M. T. Edgerton and A. R. McClary, "Augmentation Mammaplasty: Psychiatric Implications and Surgical Indications," *Plastic and Reconstructive Surgery* 21, 4 (1958): 279–305.
31. Ibid., 279.
32. Ibid., 297.
33. Ibid., 297.
34. Ibid., 297.
35. Ibid., 298.
36. Ibid., 298.
37. Ibid., 301.
38. M. T. Edgerton, E. Meyer, and W. E. Jacobson, "Augmentation Mammaplasty II: Further Surgical and Psychiatric Evaluation," *Plastic and Reconstructive Surgery* 27, 3 (1961): 300–301.
39. Julien Reich, "The Surgical Improvement in Appearance of the Female Body," *Medical Journal of Australia* 2 (November 23, 1974): 771.

40. Richard G. Druss, "Changes in Body Image Following Augmentation Breast Surgery," *International Journal of Psychoanalytic Psychotherapy* 2 (1973): 250.
41. Reich., 771.
42. James L. Baker, Irving S. Kolin, and Edmund S. Bartlett, "Psychosexual Dynamics of Patients Undergoing Mammary Augmentation," *Plastic and Reconstructive Surgery* 13, 6 (1974): 656.
43. Ibid., 658.
44. Hoopes and Knorr, "Flat-Chested Woman," 147.
45. Harvey A. Zarem, "Commentary," to Bernard L. Kaye, "Micromastia v. Macromastia," *Medical Aspects of Human Sexuality* 7 (August 1973): 120.
46. Sanford Gifford, "Emotional Attitudes toward Cosmetic Breast Surgery: Loss and Restitution of the 'Ideal Self,'" in *Plastic and Reconstructive Surgery of the Breast*, edited by Robert M. Goldwyn (Boston: Little, Brown, 1976), 107.
47. Gilbert Snyder, "Comment," to Simon Fredricks, "Managment of Mammary Hypoplasia," in *Plastic and Reconstructive Surgery of the Breast*, 408.
48. A. Richard Grossman, "Psychological and Psychosexual Aspects of Augmentation Mammaplasty," in *Symposium on Breast Surgery, Clinics in Plastic Surgery* 3, 2 (1976): 170.
49. Gifford, "Emotional Attitudes," 105.
50. Ibid., 107.
51. Patrick Clarkson and David Stafford-Clark, "Role of the Plastic Surgeon and Psychiatrist in the Surgery of Appearance," *British Medical Journal* 2 (December 17, 1960): 1768.
52. This suggestion was made by both plastic surgeons and psychiatrists. See, for example, Julien Reich, "The Surgery of Appearance: Psychological and Related Aspects," *Medical Journal of Australia* (July 5, 1969): 5–13; Richard G. Druss, Francis C. Symonds, and George F. Crikelair, "The Problem of Somatic Delusions in Patients Seeking Cosmetic Surgery," *Plastic and Reconstructive Surgery* 48, 3 (1971): 246–250.
53. For example, see Harold I. Harris, "Research in Plastic Implants," *Journal of the International College of Surgeons* 35, 5 (1961): 630–643; and Thomas J. Baker, "Cosmetic Surgery for Small Breasts," *American Journal of Nursing* 61 (June 1961): 77–78.
54. Patrick Clarkson and John Jeffs, "Modern Mammaplasty," *British Journal of Plastic Surgery* 20, 3 (1967): 307.
55. Ibid., 308.
56. Robert M. Goldwyn, "Preface," in *Plastic and Reconstructive Surgery of the Breast*, vii.
57. John Q. Owsley and Rex A. Peterson, "Preface," in *Symposium on Aesthetic Surgery of the Breast*, edited by John Q. Owsley and Rex A. Peterson (St. Louis: C. V. Mosby, 1978), ix.
58. Goldwyn, "Preface," viii.
59. Reich, "The Surgical Improvement in Appearance of the Female Body," 773.
60. Kaye, "Micromastia v. Macromastia," 96.
61. Robert H. Shipley, John M. O'Connell, and Karl F. Bader, "Personality Characteristics of Women Seeking Breast Augmentation," *Plastic and Reconstructive Surgery* 60, 3 (1977): 369–376.
62. Ibid., 370.
63. Ibid., 375.
64. Gifford, "Emotional Attitudes," 111.
65. Kaye, "Micromastia v. Macromastia," 102.
66. Grossman, "Psychological and Psychosexual Aspects of Augmentation Mammaplasty," 168.
67. Ibid., 169.
68. See, for example, Flemming Sihm, Marianne Jagd, and Michael Pers, "Psychological Assessment before and after Augmentation Mammaplasty," *Scandinavian Journal of Plastic and Reconstructive Surgery* 12 (1978): 295–298; Lennart Ohlsen, Bengt Ponten, and Gunnar Hambert, "Augmentation Mammaplasty: A Surgical and Psychiatric Evaluation of the Results," *Annals of Plastic Surgery* 2, 1 (1979): 42–52; Solveig Beale, Hans-Olof Lisper, and Bodil Palm, "A Psychological Study of Patients Seeking Augmentation Mammaplasty," *British Journal of Psychiatry* 136 (1980): 133–138.

69. L. Ombrédanne, "Restauration autoplastique du sein," *La Tribune Médicale* 38 (1906): 325–326.
70. Harold I. Harris, "Automammaplasty," *Journal of the International College of Surgeons* 12, 6 (1949): 828.
71. W. G. Holdsworth, "A Method of Reconstructing the Breast," *British Journal of Plastic Surgery* 9 (1957): 161. Holdsworth himself did perform reconstructions. In the passage quoted he was describing the objections of his colleagues.
72. Charles Horton, Jerome Adamson, and Richard Mladic, "Breast Reconstruction after Surgery and Trauma," in *Symposium on Aesthetic Surgery of the Face, Eyelid, and Breast*, 209.
73. Harold Gillies, "Surgical Replacement of the Breast," *Proceedings of the Royal Society of Medicine* 52 (1959): 599.
74. Pangman and Wallace, "Use of Plastic Prostheses," 510.
75. Richard Renneker and Max Cutler, "Psychological Problems of Adjustment to Cancer of the Breast," *JAMA* 148, 10 (1952): 833–838.
76. Ibid., 834.
77. Ibid.
78. Ibid.
79. Ibid., 835.
80. Roberta Klein, "A Crisis to Grow On," *Cancer* 28, 6 (1971): 1660–1665.
81. Michael J. Asken, "Psychoemotional Aspects of Mastectomy: A Review of Recent Literature," *American Journal of Psychiatry* 132, 1 (1975): 56.
82. Janet Polivy, "Psychological Effects of Mastectomy on a Woman's Feminine Self-Concept," *Journal of Nervous and Mental Disease* 164, 2 (1977): 118.
83. Ibid., 86.
84. Kay R. Jamison, David K. Wellisch, and Robert O. Pasnau, "Psychosocial Aspects of Mastectomy: I. The Woman's Perspective," *American Journal of Psychiatry* 135, 4 (1978): 435.
85. John Hueston and Graham McKensie, "Breast Reconstruction after Radical Mastectomy," *Australia New Zealand Journal of Surgery* 39, 4 (1970): 367.
86. Ibid.
87. John N. Simons, "Nonaesthetic Augmentation Mammaplasty," in *Symposium on Aesthetic Surgery of the Face, Eyelid, and Breast*, 173.
88. Hueston and McKensie, "Breast Reconstruction," 367.
89. Bahman Teimourian and Mehdi N. Adham, "Survey of Patients' Responses to Breast Reconstruction," *Annals of Plastic Surgery* 9, 4 (1982): 324.
90. Reuven K. Snyderman, "Reconstruction of the Breast after Surgery for Malignancy," in *Plastic and Reconstructive Surgery of the Breast*, 466.
91. Robert M. Goldwyn in "Commentary," to ibid., 484.
92. Snyderman, "Reconstruction."
93. Robert M. Goldwyn, "Commentary," 480.
94. Leonard Rubin, "Foreword," in *Symposium on Reconstruction of the Breast after Mastectomy, Clinics in Plastic Surgery* 6, 1 (1979): 1.
95. Ibid., 2.
96. Louis J. Lester, "A Critical Viewpoint by a General Surgeon toward Reconstructive Surgery after Mastectomy," in ibid., 15.
97. Snyderman, "Reconstruction," 466.
98. D. Ralph Millard, "Breast Reconstruction after a Radical Mastectomy," *Plastic and Reconstructive Surgery* 58, 3 (1976): 285.
99. Gordon Francis Schwartz, "Breast Reconstruction Following Mastectomy for Malignant Disease: A Surgical Oncologist's Point of View," in *Symposium on Reconstruction of the Breast after Mastectomy*, 5.
100. Bromley S. Freeman, "Experiences in Reconstruction of the Breast after Mastectomy," in *Symposium on Breast Surgery*, 277.
101. Vincent R. Pennisi, "Timing of Breast Reconstruction after Mastectomy," in *Symposium on Reconstruction of the Breast after Mastectomy*, 31.
102. See, for example, Gilbert Cant and Toby Cohen, "The Operation Women Never Dreamed Would Be Possible," *Good Housekeeping*, September 1975, 56+; Margaret Markham and Toby Cohen, "Miracle of a New Breast," *Harper's Bazaar*, September 1976, 148+;

"In Her Own Words: Jean Zalon Argues the Case for Breast Reconstruction after Mastectomy: 'Now I Am Whole Again,'" *People*, September 11, 1978, 69+.

103. Wendy S. Schain, Ellen Jacobs, and David K. Wellisch, "Psychosocial Issues in Breast Reconstruction: Intrapsychic, Interpersonal, and Practical Concerns," *Symposium on Advances in Breast Reconstruction*, edited by Michael Scheflan, *Clinics in Plastic Surgery* 11, 2 (1984): 242.
104. Ibid.
105. Ibid.
106. Marcia Kraft Goin, "Psychological Reactions to Surgery of the Breast," in *Symposium on Social and Psychological Considerations in Plastic Surgery, Clinics in Plastic Surgery* 9, 3 (1982): 353.
107. Schain, Jacobs, and Wellisch, "Psychosocial Issues," 242.
108. Phyllis Goldberg, Marilyn Stolzman, and Herbert M. Goldberg, "Psychological Considerations in Breast Reconstruction," *Annals of Plastic Surgery* 13, 1 (1984): 39.
109. Willard Bartlett, "An Anatomic Substitute for the Female Breast," *Annals of Surgery* (Philadelphia) 66 (1917): 208.
110. Carl O. Rice and J. H. Strickler, "Adenomammectomy for Benign Breast Lesions," *Surgery, Gynecology and Obstetrics* 93 (1951): 759.
111. Pangman and Wallace, "Use of Plastic Prostheses," 511.
112. Tibor de Cholnoky, "Late Adverse Results Following Breast Reconstructions," *Plastic and Reconstructive Surgery* 31, 5 (1963): 445–452.
113. Survey by Professional Research Analysts cited in Michael Gurdin and Gene A. Carlin, "Complications of Breast Implantations," *Plastic and Reconstructive Surgery* 40, 6 (1967): 530–533.
114. Tibor de Cholnoky, "Augmentation Mammaplasty: Survey of Complications in 10,941 Patients by 265 Surgeons," *Plastic and Reconstructive Surgery* 45, 6 (1970): 573–577.
115. David G. Bowers and Charles B. Radlauer, "Breast Cancer after Prophylactic Subcutaneous Mastectomies and Reconstruction with Silastic Prostheses," *Plastic and Reconstructive Surgery* 44, 6 (1969): 541–544.
116. Paule C. Regnault, "Indications for Breast Augmentations," *Plastic and Reconstructive Surgery* 40, 6 (1967): 524–529.
117. D. L. Weiner et al., *Subcutaneous Mastectomy with Prosthetic Reconstruction* (Midland, Mich.: Dow Corning, 1973).
118. Weiner et al. quoted in Reuven Snyderman, "'Subcutaneous Mastectomy with Immediate Prosthetic Reconstruction'—an Operation in Search of Patients," *Plastic and Reconstructive Surgery* 53, 5 (1953): 583.
119. Ibid., 583.
120. Ibid., 584.
121. Erle E. Peacock, "Biological Basis for Management of Benign Disease of the Breast," *Plastic and Reconstructive Surgery* 55, 1 (1975): 14.
122. Vincent R. Pennisi and Angelo Capozzi, "The Incidence of Obscure Carcinoma in Subcutaneous Mastectomy," *Plastic and Reconstructive Surgery* 56, 1 (1975): 9.
123. Ibid., 11.
124. Vincent R. Pennisi, "Subcutaneous Mastectomy and Fibrocystic Disease of the Breast," in *Symposium on Breast Surgery*, 210.
125. Robert M. Goldwyn, "Subcutaneous Mastectomy," *New England Journal of Medicine* 297, 9 (1977): 503.
126. Ibid., 504.
127. Anne B. Redfern and John E. Hoopes, "Subcutaneous Mastectomy: A Plea for Conservatism," *Plastic and Reconstructive Surgery* 62, 5 (1978): 706.
128. Ibid., 707.
129. Ibid., 706.
130. Reuven Snyderman, "Subcutaneous Mastectomy," in *Symposium on Aesthetic Surgery of the Breast*, 149.
131. Vincent R. Pennisi, "Subcutaneous Mastectomy," in *Long Term Results in Plastic Surgery*, II, edited by Robert M. Goldwyn (Boston: Little, Brown, 1980), 742.
132. "Breast Surgery to *Prevent* Cancer: The Big Dispute," *Good Housekeeping*, March 1981, 251.

133. Robert M. Goldwyn and Leon D. Goldman, "Subcutaneous Mastectomy and Breast Replacement," in *Plastic and Reconstructive Surgery of the Breast*, 447.
134. Loren J. Humphrey, "Subcutaneous Mastectomy Is Not a Prophylaxis against Carcinoma of the Breast: Opinion or Knowledge?" *American Journal of Surgery* 145 (March 1983): 312
135. C. Lawrence Slade, "Subcutaneous Mastectomy: Acute Complications and Long-Term Follow-up," *Plastic and Reconstructive Surgery* 73, 1 (1984): 87.
136. Jack C. Fisher, "Discussion," *Plastic and Reconstructive Surgery* 73, 1 (1984): 90.
137. Vincent R. Pennisi, "Breast Cancer and Subcutaneous Mastectomy: Facts and Opinion" (letter to the editor), *Plastic and Reconstructive Surgery* 74, 1 (1984): 153.
138. Ibid.
139. Vincent R. Pennisi and Angelo Capozzi, "Subcutaneous Mastectomy: An Interim Report on 1,244 Patients," *Annals of Plastic Surgery* 12, 4 (1984): 344.
140. Loren J. Humphrey, "Invited Comment," *Annals of Plastic Surgery* 12, 4 (1984): 345.
141. Susan M. Love, Rebecca Sue Gelman, and William Silen, "Fibrocystic 'Disease' of the Breast—a Nondisease?" *New England Journal of Medicine* 307, 16 (1982): 1010–1014.
142. Hospital discharge summary data do not indicate the reason for the procedure, but it seems fair to assume that some proportion of these procedures were for cancer prophylaxis.
143. Vincent R. Pennisi and Angelo Capozzi, "Subcutaneous Mastectomy Data: A Final Statistical Analysis of 1500 Patients," *Aesthetic Plastic Surgery* 13 (1989): 15–16.
144. Ibid., 20.
145. Ibid., 20.
146. Ibid., 20.
147. Ibid., 15.
148. Ibid., 20.
149. A 1997 study estimated that women carrying the breast cancer gene who opted for prophylactic mastectomy would gain between 2.9 to 5.3 years of life expectancy, characterized as a "substantial" gain. As the authors make clear, however, this estimate was derived not from any hard data on the effectiveness of the surgery but from consultation with "experienced clinicians." (Deborah Schrag et al., "Decision Analysis—Effects of Prophylactic Mastectomy and Oophorectomy on Life Expectancy among Women with BRCA1 or BRCA2 Mutations," *New England Journal of Medicine* 336, 20 [1997]: 1465–1471.) A 1999 report used retrospective data from women who had undergone prophylactic mastectomy between 1960 and 1993 at the Mayo Clinic. The researchers found that for those women at moderate or high risk of breast cancer, based on family history, the risk of the disease was reduced by 90 percent. (Lynn C. Hartmann et al, "Efficacy of Bilateral Prophylactic Mastectomy in Women with a Family History of Breast Cancer," *New England Journal of Medicine* 340, 2 [1999]: 77–84.)
150. This opportune pun appears in a patient education brochure produced by the ASPRS. American Society of Plastic and Reconstructive Surgeons, "Straight Talk . . . about Breast Implants," 1990.
151. "General and Plastic Surgery Devices; Effect Date of Requirement for Premarket Approval of Silicone Gel–Filled Breast Prosthesis," *Federal Register* 55, 96 (May 17, 1990): 20572.
152. Bailey Lipscomb (director of Clinical and Regulatory Affairs for Dow Corning) in transcript of the FDA's General and Plastic Surgery Devices Panel Meeting, November 12–14, 1991. Hereafter cited as November Meeting.
153. Nada Stotland (psychiatrist) in ibid.
154. Laurie Saltz in ibid.
155. Citizen Petition from the ASPRS to the FDA, November 20, 1991 (unpublished document), 5.
156. Timmie Jean Lindsay in November Meeting.
157. Bailey Lipscomb in November Meeting.
158. Citizen Petition, 21.

CHAPTER 5 REAL SUFFERING: THE CREATION OF SILICONE DISEASE

1. Karl-Heinrich Krohn, "Über Paraffinome der Mamma," *Zentralblatt für Chirurgie* 45 (1930): 2772–2781.
2. John H. Grindlay and John M. Waugh, "Plastic Sponge Which Acts as a Framework for Living Tissue," *A.M.A. Archives of Surgery* 63 (1951): 288–297.
3. Colette Perras, "The Prevention and Treatment of Infections Following Breast Implants," *Plastic and Reconstructive Surgery* 35, 6 (1965): 649.
4. R. De. R. Barondes et al., "The Silicones in Medicine," *Military Surgeon* 106 (1950): 384.
5. Earl L. Warrick, *Forty Years of Firsts: The Recollections of a Dow Corning Pioneer* (New York: McGraw-Hill, 1990).
6. "General and Plastic Surgery Device; General Provisions and Classification of 54 Devices," *Federal Register* 47, 12 (1982): 2821.
7. Dennis M. Deapen et al., "The Relationship between Breast Cancer and Augmentation Mammaplasty: An Epidemiologic Study," *Plastic and Reconstructive Surgery* 77, 3 (1986): 361–367.
8. Lynn Rosenberg, "Discussion," *Plastic and Reconstructive Surgery* 77, 3 (1986): 368.
9. Hans Berkel, Dale C. Birdsell, and Heather Jenkins, "Breast Augmentation: A Risk Factor for Breast Cancer?" *New England Journal of Medicine* 326, 25 (1992): 1649–1653.
10. Dennis M. Deapen and Garry S. Brody, "Augmentation Mammaplasty and Breast Cancer: A 5-Year Update of the Los Angeles Study," *Plastic and Reconstructive Surgery* 89, 4 (1992): 660–665.
11. Harry Hayes, James Vandergrift, and Wilma Diner, "Mammography and Breast Implants," *Plastic and Reconstructive Surgery* 82, 1 (1988): 1–6.
12. A. E. Rintala and U. M. Svinhufvud, "Effect of Augmentation Mammaplasty on Mammography and Thermography," *Plastic and Reconstructive Surgery* 54 (1974): 390. Quoted in ibid., 1.
13. For example, an article in *Glamour* reported that mammography "is not blocked and, in fact . . . it is easier to read and interpret" (325). John A. Grossman and Ruth Winter, "Breast Augmentation Surgery: How It Is Done, What You Can Expect," *Glamour*, November 1981, 255+.
14. Melvin J. Silverstein et al., "Breast Cancer in Women After Augmentation Mammaplasty," *Archives of Surgery* 123 (June 1988): 681–685.
15. K. Miyoshi et al., "Hypergammaglobulinemia by Prolonged Adjuvanticity in Man: Disorders Developed after Augmentation Mammaplasty," *Ijishimpo* (1964): 9–14. Cited in Yasuo Kumagai et al., "Clinical Spectrum of Connective Tissue Disease after Cosmetic Surgery," *Arthritis and Rheumatism* 27, 1 (1984): 1–12.
16. Cited in Thomas J. Sergott et al., "Human Adjuvant Disease, Possible Autoimmune Disease after Silicone Implantation: A Review of the Literature, Cases Studies, and Speculation for the Future," *Plastic and Reconstructive Surgery* 78, 1 (1986): 104–105.
17. K. Yoshida, "Post Mammaplasty Disorder as an Adjuvant Disease of Man," *Shikoku Acta Medica* 29 (1973): 318–332. Cited in Sergott et al., "Human Adjuvant Disease" 105.
18. Yasuo Kumagai, Chiyuki Abe, and Yuichi Shiokawa, "Scleroderma after Cosmetic Surgery," *Arthritis and Rheumatism* 22, 5 (1979): 532–537.
19. Sheryl A. van Nunen, Paul A. Gatenby, and Antony Basten, "Post-Mammaplasty Connective Tissue Disease," *Arthritis and Rheumatism* 25, 6 (1982): 694–697.
20. Testimony of Frank Vasey in *Weiss Hearing* (full citation in chapter 1, note 3), 81.
21. Kumagai et al., "Clinical Spectrum of Connective Tissue Disease."
22. John A. Byrne, *Informed Consent* (New York: McGraw-Hill, 1996).
23. Sergott et al., "Human Adjuvant Disease," 108.
24. These hypotheses are reviewed in ibid.; van Nunen, Gatenby, and Basten, "Post-Mammaplasty Connective Tissue Disease"; and Nir Kossovsky, John P. Heggers, and Martin C. Robson, "Experimental Demonstration of the Immunogenicity of Silicone Protein Complexes," *Journal of Biomedical Materials Research* 21 (1987): 1125–1133.

25. Sergott et al., "Human Adjuvant Disease," 110.
26. "General and Plastic Surgery Devices: Effect Date of Requirement for Premarket Approval of Silicone Gel–Filled Breast Prothesis," *Federal Register*, 25, 96 (May 17, 1990): 20572.
27. "Silicone Breast Implants," FDA Talk Paper, April 10, 1991; see also the transcript of "Silicone in Medical Devices," February 1–2, 1991, Baltimore, Maryland.
28. E. Jane McCarthy, Ruth B. Merkatz, and Grant Bagley, "A Descriptive Analysis of Physical Complaints from Women with Silicone Breast Implants," *Journal of Women's Health* 2, 2 (1993): 111–115.
29. Ruth B. Merkatz, Grant P. Bagley, and E. Jane McCarthy, "A Qualitative Analysis of Self-Reported Experiences among Women Encountering Difficulties with Silicone Breast Implants," *Journal of Women's Health* 2, 2 (1993): 107.
30. Patti Scher and Marion Koch, *The Untold Truth* (Charlotte, N.C.: 1993), 24.
31. Cindy Fuchs-Morrisey, "Breast Implants: A Road Map and the Choice is Yours," *AS-IS Newsletter*, January 1995: 1–3.
32. In their analysis of the part played by the media in the silicone breast implant controversy, Vanderford and Smith note that for many women "the news media provided an unexpected connection between their health problems and the silicone implants." (Marsha L. Vanderford and David H. Smith, *The Silicone Breast Implant Story: Communication and Uncertainty* [Mahwah, N.J.: Lawrence Erlbaum Associates, 1996], 41.)
33. Ibid., 116.
34. Ibid., 117.
35. Ronald E. Everson, "National Survey Shows Overwhelming Satisfaction with Breast Implants" (letter to the editor), *Plastic and Reconstructive Surgery* 88, 3 (1991): 546.
36. Neal Handel et al., "Knowledge, Concern, and Satisfaction among Augmentation Mammaplasty Patients," *Annals of Plastic Surgery* 30, 1 (1993): 13.
37. Mary Palcheff-Wiemer et al., "The Impact of the Media on Women with Implants," *Plastic and Reconstructive Surgery* 92, 5 (1993): 779–785.
38. Ibid.
39. Handel et al., "Knowledge, Concern, and Satisfaction."
40. Eric P. Winer et al., "Silicone Controversy: A Survey of Women with Breast Cancer and Silicone Implants," *Journal of the National Cancer Institute* 85, 17 (1993): 1407–1411.
41. Cited in Palcheff-Wiemer et al., "The Impact of the Media."
42. See Roberta Altman, *Waking Up/Fighting Back: The Politics of Breast Cancer* (Boston: Little, Brown, 1996) for an account of the politicization of breast cancer.
43. Tiffany Devitt, "Silicone Breast Implants" Women's Health Nightmare or 'P.R. Nightmare'?" *FAIR* (Fairness and Accuracy in Reporting), special issue (1992): 25–26.
44. For example, such case reports are an integral part of what Brown and Mikkelsen have dubbed "popular epidemiology"—"the process by which laypersons gather scientific data and other information and direct and marshal the knowledge and resources of experts to understand the epidemiology of disease." (125). (Phil Brown and Edwin J. Mikkelsen, *No Safe Place: Toxic Waste, Leukemia, and Community Action* [Berkeley: University of California Press, 1990].)
45. Testimony of Sybil Niden Goldrich in *Weiss Hearing*.
46. Sharon Green in transcript of the FDA's General and Plastic Surgery Devices Panel Meeting, November 12–14, 1991.
47. Cindy Pearson (National Women's Health Network) in ibid.
48. Joyce Ward in ibid.
49. Norman Anderson in ibid.
50. David Kessler, "Statement on Silicone Gel Breast Implants," January 6, 1992 (unpublished document).
51. ASPRS, "An Evaluation by Plastic Surgery of the 90 Documents Alleged by Commissioner David Kessler to Justify Moratorium on the Use of Silicone Gel Implants," Plastic Surgery Educational Foundation, February 1992 (unpublished document).
52. Mary Beth Jacobs (FDA staffer) in transcript of the FDA's General and Plastic Surgery Devices Panel Meeting, February 18–20, 1992. Hereafter cited as February Meeting.

53. Roselie Bright (FDA staffer) in ibid. The study upon which she based her remarks was later published: Roselie A. Bright, Lana L. Jeng, and Roscoe M. Moore, Jr., "National Survey of Self-Reported Breast Implant Estimates," *Journal of Long-Term Effects of Medical Implants* 3, 1 (1993): 81–89.
54. Citizen Petition from the ASPRS to the FDA, November 20, 1991 (unpublished document), 30.
55. Ibid., 91.
56. See, for example, the section on disease causation in Judith S. Mausner and Shira Kramer, *Mausner and Bahn Epidemiology—an Introductory Text*, 2d Ed. (Philadelphia: W. B. Saunders, 1985).
57. Jack Fisher (plastic surgeon) in February Meeting.
58. This argument is explored at length in *Phantom Risk: Scientific Inference and the Law*, edited by Kenneth R. Foster, David E. Bernstein, and Peter W. Huber (Cambridge, Mass.: MIT Press, 1993), a book which was heavily favored by conservative Republications seeking justification for tort reform.
59. Howard Tobin in February Meeting.
60. Emily Martin, *Flexible Bodies: Tracking Immunity in American Culture from the Days of Polio to the Age of AIDS* (Boston: Beacon Press, 1994), xvii.
61. Steve Kroll-Smith and H. Hugh Floyd, *Bodies in Protest: Environmental Illness and the Struggle over Medical Knowledge* (New York: New York University Press, 1997); Phil Brown and Edwin J. Mikkelsen, *No Safe Place: Toxic Waste, Leukemia, and Community Action* (Berkeley: University of California Press, 1990); Jonathan Harr, *A Civil Action* (New York: Vintage Books, 1995). The latter two are accounts of the Woburn case, in which the residents of a working-class Massachusetts town eventually won a suit against the corporate polluters whose dumping of toxic waste they blamed for the high rates of childhood leukemia and other diseases that plagued the town.
62. In *Bodies in Protest*, Kroll-Smith and Floyd characterize the "practical epistemologies" of MCS sufferers as "science joined with biography" (50): "when people appropriate a language of expertise and organize their personal lives around it" (55).

CHAPTER 6 THE SPECIALTY NEAREST SCULPTURE IN THE LIVING

1. Throughout this book, the term "plastic surgeons" refers to the collectivity as represented by their professional organizations or by what seems to be a common outlook or opinion. Of course, there are always individual surgeons whose views diverge.
2. For more on the history of plastic surgery, see *Venus Envy: A History of Cosmetic Surgery* by Elizabeth Haiken (Baltimore: Johns Hopkins University Press, 1997). Haiken presents a detailed analysis of much of the material that can only be touched upon here.
3. I am more sympathetic to the plastic surgeons than this chapter, or what you have read thus far, might lead you to think. As I spent hours poring over the before-and-after photographs in journals and textbooks, I learned this: it is a short step from perceiving something as ugly to coming to believe that plastic surgery is a right and proper intervention. Ugliness is our joint construction. We are all implicated.
4. The early history of plastic surgery is recounted in Maxwell Maltz, *Evolution of Plastic Surgery* (New York: Froben Press, 1946) and Antony F. Wallace, *The Progress of Plastic Surgery: An Introductory History* (Oxford: Willem A. Meeuws, 1982).
5. H. Lyons Hunt, *Plastic Surgery of the Head, Face and Neck* (Philadelphia: Lea and Febiger, 1926), vii.
6. Blair O. Rogers, "The Development of Aesthetic Plastic Surgery: A History," *Aesthetic Plastic Surgery* 1 (1976): 3–24.
7. Hunt, *Plastic Surgery*, 29.
8. Haiken, *Venus Envy*.
9. Gustave Aufrict, "The Development of Plastic Surgery in the United States," in *Transactions of the Thirteenth Annual Meeting of the American Society of Plastic and Reconstructive Surgery* (New Orleans, 1944).
10. Frank McDowell, "History of the American Association of Plastic Surgeons, to 1963," in Francis X. Paletta, *History of the American Society of Plastic and Reconstructive Surgery* (Baltimore: Waverly Press, 1963).

11. Frederick Figi, "The History of the American Association of Plastic Surgeons," *Plastic and Reconstructive Surgery* 5, 1 (1950): 57.
12. Paletta, *History of the American Society*, 7.
13. Ibid., 8–9.
14. In 1941, the American Association of Oral and Plastic Surgeons changed its name to the American Association of Plastic Surgeons and dropped the requirement that its members have dental degrees. The same year, the Society of Plastic and Reconstructive Surgery changed its name to the American Society of Plastic and Reconstructive Surgeons (ASPRS).
15. See Figi, "History of the American Association of Plastic Surgeons," and Aufrict, "Development of Plastic Surgery."
16. Statistics gathered from Aufrict, "Development of Plastic Surgery,"; U.S. Department of Health and Human Services, *The Current and Future Supply of Physicians and Physician Specialists* (Washington, D.C.: U.S. Government Printing Office, 1980); U.S. Department of Commerce, *Statistical Abstract of the United States*, 114th ed. (Washington, D.C.: U.S. Government Printing Office, 1994).
17. Jane Sprague Zones, "The Political and Social Context of Silicone Breast Implant Use in the United States," *Journal of Long-Term Effects of Medical Implants* 1, 3 (1992): 232.
18. The data on residency programs come from the "Directory of Approved Internships and Residencies 1960," *JAMA* 174, 6 (1960); American Medical Association, "Directory of Approved Internships and Residencies, 1969–70"; American Medical Association, "Directory of Residency Training Programs, '79–'80"; American Medical Association, "Graduate Medical Education Directory, 1993–1994."
19. All of the figures on ADC and outpatient visits were collected from the Residency Directories for the relevant years.
20. John Q. Owsley and Rex A. Peterson, "Preface," in *Symposium on Aesthetic Surgery of the Breast*, edited by John Q. Owsley and Rex A. Peterson (St. Louis: C. V. Mosby, 1978), ix.
21. E. H. De Kleine, "The Crossroads of Cosmetic Surgery," *Plastic and Reconstructive Surgery* 10, 2 (1955): 145–150.
22. Zones, "Political and Social Context," 232.
23. Mary Ann Eiler and Thomas J. Pasko, *Specialty Profiles, 1988 Edition* (Chicago: American Medical Association, 1988).
24. American Society of Plastic and Reconstructive Surgeons Department of Communications, "Estimated Number of Cosmetic Surgery Procedures Performed by ASPRS Members, 1988."
25. Cost figures were gathered from popular press accounts from the 1970s to 1990.
26. Christopher Drew and Michael Tacket, "Access Equals Clout: The Blitzing of FDA," *Chicago Tribune*, December 8, 1992.
27. ASPRS, "The History of Plastic Surgery," undated news release.
28. Wallace, *Progress of Plastic Surgery*, 57.
29. Frank W. Masters, "Preface," in *Symposium on Aesthetic Surgery of the Face, Eyelid, and Breast*, edited by Frank W. Masters and John R. Lewis (St. Louis: C. V. Mosby, 1972), ix.
30. Robert M. Goldwyn , "Preface," in *Plastic and Reconstructive Surgery of the Breast*, edited by Robert M. Goldwyn (Boston: Little, Brown, 1976), ix.
31. This trend is described in Paul Starr, *The Social Transformation of American Medicine* (New York: Basic Books, 1982).
32. The gendered nature of this psychological affinity is revealed by the tendency of plastic surgeons to fear male patients. Until very recently, a man who sought cosmetic surgery (especially a "nose job") was automatically suspect. Plastic surgeons were warned to exercise special care in the treatment of these patients, who were believed to be prone to violence.
33. J. Eastman Sheehan, "Foreword," to Max Thorek, *Plastic Surgery of the Breast and Abdominal Wall* (Springfield: Charles C. Thomas, 1942), ix.
34. Herbert Conway in *Transactions of the Third International Congress of Plastic Surgery* (Amsterdam: Excerpta Medica Foundation, 1964), 31.
35. Gustave Aufrict, "Philosophy of Plastic Surgery," *Plastic and Reconstructive Surgery* 20, 5 (1957): 397.

36. John Hoopes and Norman Knorr, "Psychology of the Flat-Chested Woman," in *Symposium on Aesthetic Surgery of the Face, Eyelid, and Breast*, 148.
37. Paul P. Pickering, "Socioeconomic Considerations of Aesthetic Surgery," in *Symposium on Aesthetic Surgery of the Face, Eyelid, and Breast.*
38. Hoopes and Knorr, "Psychology of the Flat-Chested Woman," 148.
39. Frederick Strange Kolle, *Plastic and Reconstructive Surgery* (New York, D. Appleton, 1911), 214.
40. This subtext is made explicit in James L. Baker, Irving S. Kolin, and Edmund S. Bartlett, "Psychosexual Dynamics of Patients Undergoing Mammary Augmentation," *Plastic and Reconstructive Surgery* 13, 6 (1974): 652–659.
41. Sanford Gifford, "Emotional Attitudes toward Cosmetic Breast Surgery: Loss and Restitution of the 'Ideal Self,'" in *Plastic and Reconstructive Surgery of the Breast.*
42. Richard G. Druss, "Changes in Body Image Following Augmentation Breast Surgery," *International Journal of Psychoanalytic Psychotherapy* 2 (1973): 247.
43. Gifford, "Emotional Attitudes," 113.
44. See, for example, David Maddison, "Augmentation Mammaplasty: A Psychiatrist's View," *Australia and New Zealand Journal of Surgery* 46, 4 (1976): 355–358.
45. John A. Grossman and Ruth Winter, "Breast Augmentation Surgery: How It Is Done, What You Can Expect," *Glamour*, November 1981, 317.
46. See, for example, Arthur Frank and Stuart Frank, "Cosmetic Breast Surgery," *Mademoiselle*, November 1975, 74+.
47. "The More the Mariel," *People*, November 21, 1983, 40.
48. Joanne Kaufman, "Whose Breasts Are They, Anyway," *Mademoiselle*, August 1987, 70.
49. "Yes, You Can Have a Bigger Bosom!" *Cosmopolitan*, January 1970, 66.
50. Judith Ramsey, "Healthier Bosom: All the Latest Medical and Surgical News," *Ladies Home Journal*, October 1969, 82.
51. Harriet LaBarre, "Making More (or Less) of Your Bosom," *Ladies Home Journal*, May 1972, 127.
52. Sharon Romm, "Who Is a Good Candidate for Cosmetic Surgery?" *Vogue*, October 1982, 182.
53. Mark Gorney, "Everything You Ever Wanted to Know About Malpractice But Were Afraid to Ask," undated, unpaginated document distributed by the ASPRS.
54. Baker et al., "Psychosexual Dynamics," 658.
55. Starr, *Social Transformation*, 23.
56. Aufrict, "Philosophy of Cosmetic Surgery," 397.
57. Robby Meijer, "When Your Patient Asks about Plastic Surgery," *Medical Times* 99, 9 (September 1971): 117.
58. Michael Scheflan, "Foreword," in *Symposium on Advances in Breast Reconstruction*, edited by Michael Scheflan, *Clinics in Plastic Surgery* 11, 2 (1984): 229.
59. For example, in the textbook *Plastic and Reconstructive Surgery of the Breast*, Garry S. Brody wrote that "Homo sapiens is the only species that modifies physical appearance by decoration. The variety of decor in the human experience is almost infinite because of our unique ability to reason, imagine, and prognosticate, which is facilitated by tool-making skills." He goes on to describe scarification, foot binding, and ear piercing as examples of physical "decor," asking, "Can face lifts, liposuction, or breast implants be any more civilized?" (119).
60. John E. Alexander, "Challenges in Esthetic Plastic Surgery," *Plastic and Reconstructive Surgery* 52, 4 (1973): 337.
61. Colette Perras, "Foreword," in *Symposium on Breast Surgery*, edited by Colette Perras, *Clinics in Plastic Surgery* 3, 2 (1976): 165.
62. M. Gonzalez-Ulloa, "History of Rhytidectomy," in *The Creation of Aesthetic Plastic Surgery*, edited by M. Gonzalez-Ulloa (New York: Springer-Verlag, 1985), 81.
63. John F. Crosby, "Aesthetics and the Aesthetic Plastic Surgeon," in *The Art of Aesthetic Plastic Surgery*, edited by John R. Lewis (Boston: Little, Brown, 1989), 3.
64. In emphasizing the importance of claims, I draw upon the social construction of problems approach, which views claims and claims making as the central activities of meaning making. For explorations of the dynamics of claims, see Joseph R. Gusfield, *The Culture of Public Problems: Drinking-Driving and the Symbolic Order* (Chicago:

University of Chicago Press, 1981); Malcolm Spector and John I. Kitsuse, *Constructing Social Problems* (New York: Aldine De Gruyter, 1987); and Stephen Hilgartner and Charles S. Bosk, "The Rise and Fall of Social Problems: A Public Arenas Model," *American Journal of Sociology* 94, 1 (July 1988): 53–78.

65. "General and Plastic Surgery Device; General Provisions and Classification of 54 Devices," *Federal Register* 47, 12 (January 19, 1982): 2820.
66. "Comments of the American Society of Plastic and Reconstructive Surgeons on the Proposed Classification of Inflatable Breast Prosthesis (Docket No. 78N-2653) and Silicone Gel–Filled Breast Prosthesis (Docket No. 78N-2654); 47 *Fed. Reg.* 2810, January 19, 1982," submitted to the FDA July 1, 1982 (unpublished document), 4–5.
67. Mary H. McGrath and Boyd R. Burkhardt, "The Safety and Efficacy of Breast Implants for Augmentation Mammaplasty," *Plastic and Reconstructive Surgery* 74, 4 (1984): 550–560.
68. Ibid., 550.
69. Ibid., 550.
70. Ibid., 551.
71. Ibid.
72. Ibid.
73. The difference between efficacy and effectiveness is not one of semantics. Efficacy refers to whether an intervention works under ideal circumstances, whereas effectiveness denotes performance in the real world. Because the plastic surgeons wanted to emphasize that the FDA had no authority over physician practice—the site of effectiveness—it was in their interest to emphasize the efficacy criterion.
74. ASPRS, "Statement of the American Society of Plastic and Reconstructive Surgeons, Submitted to the House Committee on Government Operations, Subcommittee on Human Resources and Intergovernmental Relations by Garry S. Brody, RE: Silicone Breast Implants, for the Hearing of December 18, 1990."
75. ASPRS, "Straight Talk . . . about Breast Implants," 1990. Reproduced in the transcript of the *Weiss Hearing* (full citation in chapter 1, note 3).
76. Letter from Bruce Williams and Norman Cole to ASPRS membership, 1991.
77. "Congress Asked to Help Preserve Women's Right to Choose Implants," ASPRS news release, October 8, 1991.
78. ASPRS, "Key Message Points for Physicians, Nurses and Staff Letters to FDA and Congress," fall 1991 (unpublished document).
79. Letter from Bruce Williams and Norman Cole to ASPRS membership, 1991.
80. ASPRS, "Media Tips," *Breast Implant Bulletin*, October 25, 1991.
81. ASPRS president Norman Cole quoted in ASPRS news release of October 8, 1991.
82. ASPRS, "Media Tips."
83. ASPRS, "Key Message Points."
84. These arguments were advanced in various ASPRS documents and in testimony to FDA advisory panels.
85. Lori Saltz (female plastic surgeon) quoted in ASPRS news release of October 8, 1991.
86. ASPRS, "Sample Letter from Physicians to Patients," fall 1991 (unpublished document).
87. Ibid.
88. For example, the Boston Women's Health Book Collective distributed a critique of the ad, charging that it contained "false and misleading statements" and arguing that the "issue is safety, not freedom of choice." The collective urged women to "write to the FDA immediately to urge it to make a decision based on the facts, not politics." ("What's Wrong with the Ad from the Plastic Surgeons?" Boston Women's Health Book Collective, 1991.)
89. As Phil Brown and Edwin J. Mikkelsen explore in their account of the Woburn case, the tendency to psychologize complaints in order to dismiss them is a common strategy in toxic torts and other injury lawsuits. (Phil Brown and Edwin J. Mikkelsen, *No Safe Place: Toxic Waste, Leukemia, and Community Action* [Berkeley: University of California Press, 1990].)
90. Several years later, of course, tobacco would itself be at the center of a similar controversy involving "victims," industry, and the FDA's power to regulate.

91. Thomas DeWire in transcript of the FDA's General and Plastic Surgery Devices Panel Meeting, November 12–14, 1991.
92. Laurie[sic] Saltz in ibid.
93. Citizen Petition from the ASPRS to the FDA, November 20, 1991 (unpublished document), 4.
94. Ibid., 107–108.
95. "An Evaluation by Plastic Surgery of the 90 Documents Alleged by Commissioner David Kessler to Justify Moratorium on the Use of Silicone Gel Implants," prepared by members of the Silicone Implant Research Committee, ASPRS Plastic Surgery Educational Foundation, February 1992 (unpublished document), 15.
96. Garry Brody in transcript of the FDA's General and Plastic Surgery Devices Panel Meeting, February 18–20, 1992.
97. James Baker in ibid.
98. Howard Tobin in ibid.

CHAPTER 7 A CALCULUS OF RISK

1. Harvey Wiley's role is described by Wallace F. Janssen, "Introduction," *Annual Reports, 1950–1974 on the Administration of the Federal Food, Drug, and Cosmetic Act and Related Laws* (Washington, D.C.: U.S. Government Printing Office, 1976).
2. *A Brief Legislative History of the Food, Drug, and Cosmetic Act, Prepared by the Staff for the Use of the Committee on Interstate and Foreign Commerce and Its Subcommittee on Public Health and the Environment, U.S. House of Representatives* (Washington, D.C.: U.S. Government Printing Office, 1974).
3. Henry A. Lepper, "Fifty Years of Protection under Federal Food and Drug Laws," in *The Impact of the Food and Drug Administration on Our Society*, edited by Henry Welch and Felix Marti-Ibanez (New York: MD Publications, 1956).
4. Janssen, "Introduction."
5. *A Brief Legislative History*, 2–3.
6. Ibid., 16.
7. *Medical Devices Legislation, Prepared by the Staff for the Use of the Subcommittee on Health and the Environment of the Committee on Interstate and Foreign Commerce. U.S. House of Representatives* (Washington, D.C.: U.S. Government Printing Office, 1975).
8. The history of the Cooper Committee is reviewed in ibid.
9. Herbert Burkholz, *The FDA Follies* (New York: Basic Books, 1994).
10. Ibid., 5–6.
11. George P. Larrick, "The Food and Drug Administration of Today," in *Impact of the Food and Drug Administration on Our Society*, 15.
12. Robert A. Hardt, "A Pharmaceutical Manufacturer Looks at the FDA" in ibid., 56.
13. Charles Wesley Dunn, "A Practicing Attorney Looks at the 1938 Food, Drug, and Cosmetic Act and Its Administration" in ibid., 101.
14. Susan G. Hadden, "DES and the Assessment of Risk," in *Controversy: Politics of Technical Decisions*, 2d ed., edited by Dorothy Nelkin (Beverly Hills, Calif.: Sage Publications, 1984), 111–124.
15. *How Safe Is Safe?: The Design of Policy on Drugs and Food Additives* (Washington, D.C.: National Academy of Sciences, 1974), 48.
16. Ibid., 213.
17. John Blair quoted in Burkholz, *FDA Follies*, 13.
18. *A Brief Legislative History.*
19. Albert H. Holland, "How the Federal Food, Drug, and Cosmetic Act Helps the Physician," in *Impact of the Food and Drug Administration on Our Society.*
20. Ibid., 20.
21. William M. Wardell and Louis Lasagna, *Regulation and Drug Development* (Washington D.C.: American Enterprise Institute, 1975), 14.
22. Ibid., 15.
23. Garry S. Brody, "Overview of Augmentation Mammaplasty," in *Plastic and Reconstructive Surgery of the Breast*, edited by R. Barrett Noone (Philadelphia: B. C. Decker, 1991). Brody wrote: "The [Medical Devices] law regulates only the importation and

sale of these devices in interstate commerce, and has jurisdiction only over manufacturers and distributors. The FDA has no jurisdiction over any device that is custom made, because these are considered as being for one-time use in an individual patients. Surgeons may legally implant anything into the human body without the concern of the FDA" (122).

24. This image of the FDA as a meritocracy of crew-cut, taciturn heroes dominates a children's book about the agency: Josephine Hemphill, *Arsenic and Fruitcake* (Boston: Little, Brown, 1962). It is also apparent in FDA commissioner David Kessler's nickname—Eliot Knessler, a pun on the name of the crusading leader of the "Untouchables," federal agents who brought down Al Capone.
25. James C. Peterson and Gerald E. Markle, "The Laetrile Controversy," in *Controversy*, 175–196.
26. David J. Rothman and Harold Edgar, "New Rules for New Drugs: The Challenge of AIDS to the Regulatory Process," *Milbank Quarterly* 68 (supplement 1) (1990): 111–142.
27. Each of these groups is clearly involved in equally intricate relationships with each other, as well. For example, physicians are part of the pharmaceutical and device industries: they are stockholders and inventors, as well as researchers funded by industry to perform FDA-mandated studies of the safety and effectiveness of new products.
28. Robert W. Crandall and Lester B. Lave, "Introduction" and "Summary," in *The Scientific Basis of Health and Safety Regulation*, edited by Robert W. Crandall and Lester B. Lave (Washington, D.C.: Brookings Institute, 1981).
29. *Medical Device Amendments of 1976*, Public Law 94–295, 94th Cong., May 28, 1976.
30. "Classification Procedures—Final Rule," *Federal Register* 43, 146 (July 28, 1978): 32995.
31. Ibid.
32. James Turner in *How Safe Is Safe?* 14.
33. Elizabeth Jacobson, "Regulatory Science: Coping with Uncertainty," presented at the FDA-sponsored conference *Silicone in Medical Devices*, Baltimore, Maryland, 1991.
34. "Medical Devices: Classification Procedures," *Federal Register* 43, 177 (September 13, 1977): 46036.
35. "Classification Procedures—Final Rule," 32995.
36. Ibid., 32991.
37. Henry G. Grabowski and John M. Vernon, *The Regulation of Pharmaceuticals* (Washington, D.C.: American Enterprise Institute, 1983).
38. James Turner in *How Safe Is Safe?* 14.
39. Peter Barton Hutt in ibid.
40. Analyses of public controversies over science and technology show how often uncertainty forms the basis of these disputes. See, for example, Nelkin's *Controversy*; *Science in Context*, edited by Barry Barnes and David Edge (Cambridge, Mass.: MIT Press, 1982); and *Phantom Risk: Scientific Inference and the Law*, edited by Kenneth R. Foster, David E. Bernstein, and Peter W. Huber (Cambridge, Mass.: MIT Press, 1993). These analysts describe at least three different forms of uncertainty: As Nelkin suggests, one aspect of uncertainty is "limited knowledge." In adversarial situations, factual uncertainty leads to different interpretations of the same data. Another aspect of uncertainty is described by Barnes and Edge, who note that the inductive, probablistic character of science leaves it open to challenge on epistemologic grounds. A third sort of uncertainty, the focus of *Phantom Risk*, develops when science is used in other arenas, specifically the courtroom, and becomes subject to standards of assessment designed for other kinds of evidence.
41. "Classification Procedures—Final Rule," 32991.
42. I use the concept of scientism here as it was explicated by Barnes and Edge in *Science in Context*. The "scientization of politics," or scientism, occurs when "the infiltration of technical expertise determines the *conceptualization* of political problems, the *language* in which they are expressed, and the *institutional forms* by which decisions are reached" (244). Scientism becomes a tool to maintain social order and a mechanism to limit public participation in social life to those with a certain level of education and

expertise. In public controversies, like the breast implant issue, scientism structures the form and content of evidence and argument.

43. Crandall and Lave, *Scientific Basis*, 14.
44. Dorothy Nelkin ("Controversy as Political Challenge" in *Science in Context*) describes the blurring of "facts" (rationality) and "values" (politics) that often occurs in public controversies.
45. "General and Plastic Surgery Device; General Provisions and Classification of 54 Devices," *Federal Register* 47, 12 (January 19, 1982): 2810–2847.
46. Ibid., 2820.
47. Ibid., 2821.
48. Such debate, however, was not independent of the agency, but rather was an effect of the agency's actions. For example, the FDA located the debate in letters sent to the agency and in the content of testimony made to FDA advisory panels.
49. "Comments of the American Society of Plastic and Reconstructive Surgeons on the Proposed Classification of Inflatable Breast Prothesis (Docket No. 78N-2653) and Silicone Gel–Filled Breast Prosthesis (Docket No. 78N–2654); *Fed. Reg.* 2810, January 19, 1982," submitted to the FDA July 1, 1982 (unpublished document), 12.
50. These objections are described in the text of "General and Plastic Surgery Devices; General Provisions and Classifications of 51 Devices—Final Rule," *Federal Register* 53, 122 (June 24, 1988): 23860. Hereafter cited as "Final Rule, 1988."
51. "Comments of the American Society of Plastic and Reconstructive Surgeons on the Proposed Classification," 47–48.
52. "Final Rule, 1988," 23863.
53. The legislative language of the Medical Device Amendments mandated that Class III be the default classification for implantable devices.
54. "Final Rule, 1988," 23860.
55. Ibid.
56. Described in Patti Scher and Marion Koch, *The Untold Truth about Breast Implants* (Charlotte, N.C., 1993), 98.
57. Ibid.; see also John A. Byrne, *Informed Consent* (New York: McGraw-Hill, 1996), 119.
58. Warren E. Leary, "FDA Is Ordered to Release Data on the Safety of Breast Implants," *New York Times*, November 29, 1990.
59. Decision by U.S. District Court Judge Stanley Sporkin quoted in Scher and Koch, *Untold Truth*, 99.
60. "General and Plastic Surgery Devices; Effective Date of Requirement for Premarket Approval of Silicone Gel–Filled Breast Prosthesis," *Federal Register* 55, 96 (May 17, 1990): 20572.
61. Ibid., 20571.
62. "General and Plastic Surgery Devices; Effective Date of Requirement for Premarket Approval of Silicone Gel-filled Breast Prosthesis," *Federal Register* 56, 69 (April 10, 1991): 14625. Hereafter cited as "PMAA Requirement, 1991."
63. Ibid., 14622.
64. Ibid., 14621.
65. Martin Kasindorf, "'Quayle Council' Stirs Up Capital; Lauded by Business, but Hit by Consumerists," *Newsday*, September 9, 1991.
66. Ibid.
67. See, for example, Philip J. Hilts, "Questions on Role of Quayle Council," *New York Times*, November 19, 1991.
68. Philip J. Hilts, "Proposal Seeks to Speed Up Federal Approval of Drugs," *New York Times*, November 9, 1991.
69. Ibid.
70. Dana Priest, "Competitiveness Council Under Scrutiny; Critics Charge Panel Lets Industry Exert Back-Door Influence on Implementing Laws," *Washington Post*, November 26, 1991; Philip J. Hilts, "Concern on Plans by Quayle Council," *New York Times*, March 20, 1992; "House Panel Raps Quayle Council's FDA Plan," *Chicago Tribune*, October 12, 1992.
71. Malcolm Gladwell, "FDA Implements Changes in Drug Approval Process," *Washington Post*, April 10, 1992.
72. "PMAA Requirement, 1991."

73. "Silicone Breast Implants," FDA Talk Paper, April 10, 1991.
74. All quotes in this paragraph are taken from the opening remarks of David Kessler in transcript of the FDA's General and Plastic Surgery Devices Panel Meeting, November 12–14, 1991.
75. Daniel McGunagle (FDA staffer) in ibid.
76. Testimony of Bailey Lipsomb (Dow Corning director of Clinical and Regulatory Affairs) in ibid.
77. Dennis Condon (Mentor Corporation) in ibid. (Later, however, the silicone used in Norplant did become an issue: as women had the devices removed from the insertion site in the upper arm, many suffered scaring as a result of tissue adhering to the silicone—i.e., forming a kind of capsule around the device. Such difficult removals and resultant scarring became the basis for suits against the contraceptive's manufacturer.)
78. Ibid.
79. Ibid.
80. All quotes are from David Kessler, "Statement on Silicone Gel Breast Implants," January 6, 1992.
81. Ronald Johnson (FDA staffer) in transcript of the FDA's General and Plastic Surgery Devices Panel Meeting, February 18–20, 1992.
82. David Kessler in ibid.
83. Jack Fisher (plastic surgeon) in ibid.
84. Marilyn Lloyd in ibid.
85. David Kessler quoted in HHS news release, April 16, 1992.
86. Marcia Angell, "Breast Implants—Protection or Paternalism?" *New England Journal of Medicine* 326, 25 (1992): 1695–1696. Angell later expanded her critique into a book: *Science on Trial: The Clash of Medical Evidence and the Law in the Breast Implant Case* (New York: W. W. Norton, 1996).
87. Ibid., 1695.
88. Ibid., 1696.
89. Ibid.
90. Ibid.
91. Ibid.
92. Jack Fisher, "The Silicone Controversy—When Will Science Prevail?" *New England Journal of Medicine* 326, 25 (1992): 1696.
93. David Kessler, "The Basis of the FDA's Decision on Breast Implants," *New England Journal of Medicine* 326, 25 (1992): 1713–1715.
94. Ibid., 1714.
95. Ibid., 1715.
96. *The FDA's Regulation of Silicone Breast Implants, a Staff Report Prepared by the Human Resources and Intergovernmental Relations Subcommittee of the Committee on Government Relations, December 1992* (Washington, D.C.: U.S. Government Printing Office, 1993).

CHAPTER 8 AFTER BABEL: AN EPILOGUE

1. For example, Nir Kossovsky and Nora Papasian, "Clinical Reviews: Mammary Implants," *Journal of Applied Biomaterials* 3, 3 (1992): 239–242; Nachman Brautbar, Aristo Vojdani, and Andrew W. Campbell, "Silicone Implants and Systemic Immunological Disease: Review of the Literature and Preliminary Results," *Toxicology and Industrial Health* 8, 5 (1992): 231–237; Steven H. Yoshida et al., "Silicon and Silicone: Theoretical and Clinical Implications of Breast Implants," *Regulatory Toxicology and Pharmacology* 17 (1993): 3–18.
2. For example, Nir Kossovsky et al., "Surface Dependent Antigens Identified by High Binding Avidity of Serum Antibodies in a Subpopulation of Patients with Breast Prostheses," *Journal of Biomaterials* 4 (1993): 281–288; Marc Lappe, "Silicone-Reactive Disorder: A New Autoimmune Disease Caused by Immunostimulation and Superantigens," *Medical Hypotheses* 41 (1993): 348–352.
3. These public statements are, respectively: Garry S. Brody et al., "Consensus Statement on the Relationship of Breast Implants to Connective Tissue Disease," *Plastic and Reconstructive Surgery* 90, 6 (1992): 1102–1105; "UK Advice on Silicone Breast

Implants," *Lancet* 339 (February 1, 1992): 300; Independent Advisory Committee on Silicone-Gel-Filled Breast Implants, "Summary of the Report on Silicone-Gel-Filled Breast Implants," *Canadian Medical Association Journal* 147, 8 (1992): 1141–1146; Council on Scientific Affairs, American Medical Association, "Silicone Gel Breast Implants," *JAMA* 270, 21 (1993): 2602–2608.

4. Randall M. Goldblum et al., "Antibodies to Silicone Elastomers and Reactions to Ventriculoperitoneal Shunts," *Lancet* 340 (August 1992): 510–513.
5. Raymond I. Press et al., "Antinuclear Antibodies in Women with Silicone Breast Implants," *Lancet* 340 (November 1992): 1304–1307.
6. Teri Randall, "Less Maligned, but Cut from the Same Cloth, Other Silicone Implants Also Have Adverse Effects" and "Surgeons Grapple with Synovitis, Fractures around Silicone Implants for Hand and Wrist," *JAMA* 268, 1 (1992): 12–13, 17–18.
7. Sandra Blakeslee, "Dow Found Silicone Danger in 1975 Study, Lawyers Say," *New York Times*, April 7, 1994.
8. Richard Stone, "The Case Against Implants," *Science* 260 (April 2, 1993): 31.
9. Jeremiah J. Levine and Norman T. Ilowite, "Sclerodermalike Esophageal Disease in Children Breastfed by Mothers with Silicone Breast Implants," *JAMA* 271, 3 (1994): 213–216.
10. Gina Kolata, "Tissue Illness and Implants: No Tie Is Seen," *New York Times*, May 29, 1994.
11. Sherine E. Gabriel et al., "Risk of Connective-Tissue Diseases and Other Disorders after Breast Implantation," *New England Journal of Medicine* 330, 24 (1994): 1697–1702.
12. Jorge Sanchez-Guerrero et al., "Silicone Breast Implants and the Risk of Connective Tissue Diseases and Symptoms," *New England Journal of Medicine* 332, 25 (1995): 1666–1670.
13. Charles H. Hennekens et al., "Self-Reported Breast Implants and Connective-Tissue Diseases in Female Health Professionals," *JAMA* 275, 8 (1996): 616–621.
14. For example, an Associated Press report of the Mayo Clinic study ran in the *Baltimore Sun* (June 17, 1994) under the headline "Breast Implants Cleared by Study."
15. For example, see Gary Solomon et al., "Letter to the Editor," *New England Journal of Medicine* 332, 19 (1995): 1306–1307; Richard Alexander, "Breast Implants: Fact v. Fiction in the Harvard and Mayo Clinic Studies," Alexander Law Firm, 1995 (unpublished document obtained on the Internet).
16. Kessler made these points when he was interviewed on the *Diane Rehm Show* (NPR), November 22, 1995.
17. Barbara G. Silverman et al., "Reported Complications of Silicone Gel Breast Implants: An Epidemiologic Review," *Annals of Internal Medicine* 124, 8 (1996): 744.
18. Cori Vanchieri, "European Surgeons Call for Independence from U.S. Food and Drug Administration," *Journal of the National Cancer Institute* 85, 5 (1993): 2439.
19. Ministère des Affaires Sociales de la Santé et de la Ville, "Prothèses mammaires implantables," press release, January 24, 1995.
20. D. M. Gott and J.J.B. Tinkler, *Silicone Implants and Connective Tissue Disease*, report issued by the Medical Devices Agency, 1994.
21. Dennis M. Deapen, "Are Breast Implants Anticarcinogenic? A 14-year Follow-up of the Los Angeles Study," *Plastic and Reconstructive Surgery* 99, 5 (1997): 1346–1353.
22. For examples of such reviews, see Warren D. Blackburn, Jr. and Michael P. Everson, "Silicone-Associated Rheumatic Disease: An Unsupported Myth," *Plastic and Reconstructive Surgery* 99, 5 (1997): 1362–1367; Sheryl L. Lewin and Timothy A. Miller, "A Review of Epidemiologic Studies Analyzing the Relationship between Breast Implants and Connective Tissue Diseases," *Plastic and Reconstructive Surgery* 100, 5 (1997): 1309–1313.
23. Gina Kolata, "British Panel Disputes Risks of Implants Using Silicone," *New York Times*, July 15, 1998.
24. *The Newshour with Jim Lehrer*, June 21, 1999.
25. Shirley Frondorf, "Silicone Implants and the 'Global' settlement," *Breast Cancer Action*, April 1994.

26. "Silicone Breast Implants Didn't Cause Plaintiff's Injuries, Jury Decides," *Wall Street Journal*, August 20, 1992; "Jurors Absolve Maker of Implant," *New York Times*, June 12, 1993.
27. "3 Are Awarded $27.9 Million in Implant Trial," *New York Times*, March 4, 1994.
28. Paula F. Henry, "Overview of Breast Implant Litigation," *Nurse Practitioner Forum* 3, 4 (1992): 189–190.
29. Gina Kolata, "Fund Proposed for Settling Suits over Breast Implants," *New York Times*, September 10, 1993.
30. Gina Kolata, "Details of Implant Settlement Announced by Federal Judge," *New York Times*, April 5, 1994.
31. This summary is a distillation of the information women who responded to the notice of the settlement received when they answered the media advertisements. It is contained in the Breast Implant Litigation Settlement Agreement, U.S. District Court, Northern District of Alabama, 1994.
32. Gina Kolata, "3 Companies in Landmark Accord on Lawsuits over Breast Implants," *New York Times*, March 24, 1994.
33. *CTN Newsletter*, fall 1994.
34. John Schwartz, "Jury Faults Dow Chemical in Breast Implant Trial," *New York Times*, August 19, 1997.
35. Edward J. Burger, Jr., *Better Science by Litigators, Better Management by Courts: Drawing on the Pre-Daubert Experience in the Post-Daubert Era* (Washington, D.C.: National Legal Center for the Public Interest, 1997).
36. Michael E. Reed, "*Daubert* and the Breast Implant Litigation: How Is the Judiciary Addressing the Science?" *Plastic and Reconstructive Surgery* 100, 5 (1997): 1322–1326.
37. One such report was produced by the Washington D.C.–based Institute for Women's Policy Research. The author was Diana Zuckerman, formerly a key staffer for Congressman Ted Weiss. Zuckerman argued that all of the existing research had methodological flaws (including small sample sizes, a lack of specification of implant type, short-term follow-up, and poor comparison samples) that rendered them inadequate to answer the central question about disease causation. (Diana Zuckerman, *The Safety of Silicone Breast Implants* [Washington, D.C.: Institute for Women's Policy Research, 1998].)
38. Ralph R. Cook, Myron C. Harrrison, and Robert R. LeVier, "The Breast Implant Controversy," *Arthritis and Rheumatism* 37, 2 (1994): 153.
39. Ibid., 154.
40. Ibid., 154
41. Ibid., 153.
42. Kolata, "Tissue Illness and Implants."
43. Marsha F. Goldsmith, "Images of Perfection Once the Goal—Now Some Women Just Seek Damages," *JAMA* 267, 18 (1992): 2439.
44. Ibid.
45. The speech is recounted in "Beauty or the Beast," *Physician's Weekly*, February 1, 1993.
46. "The Breast Implant Crisis," *Plastic and Reconstructive Surgery* 94, 4 (1994): 97A.
47. Ibid., 91A.
48. Christina Slawson and Elizabeth Graham, "Breast Implants and the Law," *Journal of Law, Medicine, and Ethics* 21, 1 (1993): 122–126.
49. ASPRS, "Summary of the Development of an In-Kind Surgical Services Program," circa spring 1994 (unpublished document).
50. Ibid.
51. "The Breast Implant Crisis," 91A.
52. Ibid., 99A.
53. "Beauty or the Beast," 1993.
54. "The Breast Implant Crisis," 99A.
55. In 1997, when I was working as a health policy fellow for the Health Subcommittee of the Senate Labor and Human Resources Committee, a bill was introduced in the House that would have required insurers that covered the cost of mastectomy also to provide payment for the cost of reconstruction and for surgery on the "remaining

breast." The Reconstructive Breast Surgery Benefits Act was introduced in the House by Anna Eshoo, Democrat from California, and was supported (and heavily promoted) both by breast cancer advocacy groups and by the ASPRS. In the fall of 1998, Congress passed the legislation under the title "Women's Health and Cancer Rights Act."

56. John A. Byrne, *Informed Consent* (New York: McGraw-Hill, 1996).
57. Sheryl Gay Stolberg, "Grateful to Be Heard, 'Silicone Survivors' Take Fight over Implants to Capital," *New York Times*, July 25, 1998.
58. The history of the breast cancer advocacy movement is recounted in Karen Stabiner, *To Dance with the Devil: The New War on Breast Cancer* (New York: Delacorte Press, 1997) and Roberta Altman, *Waking Up/Fighting Back: The Politics of Breast Cancer* (Boston: Little, Brown, 1996).
59. The *60 Minutes* segment—"Women at Risk?"—aired October 22, 1995. The *Frontline* documentary, titled "Breast Implants on Trial," aired on PBS on February 27, 1996.
60. The Food and Drug Administration Modernization and Accountability Act was passed by Congress and signed by President Bill Clinton in the fall of 1997.
61. Although generally described as "safer than silicone," and not subject to the FDA's strictures on use, saline-filled breast implants have also been linked to a variety of health problems. Explanted devices have shown signs of fungal or bacterial growth. One of the first implant-related events I attended was a 1994 FDA hearing on the safety of saline implants, at which women claiming to have been harmed by these devices testified to horror stories as chilling as those reported with silicone gel–filled implants.
62. Denise Grady, "Cosmetic Breast Enlargements Making a Comeback," *New York Times*, July 21, 1998. Grady credits the ASPRS for her statistics.
63. Jane E. Allen, "Comfort and Conflict," *Los Angeles Times*, December 14, 1998. The existence of such a black market is a matter of ambivalence to plastic surgeons. On the one hand, that it appears to have sprung up supports some of the contentions about need made during the controversy; on the other, the fact that plastic surgeons in the United States are willing to do augmentation procedures using these black market devices tends to criminalize the profession.
64. Rod J. Rohrich, et al., "Development of Alternative Breast Implant Filler Material: Criteria and Horizons," *Plastic and Reconstructive Surgery* 98, 3 (1996): 552–560. These alternatives include soybean oil, hyaluronic acid, polyethylene glycol, and a saline-chemical mix called "hydrogel."
65. See, for example, Dawn Margolis, "Fat Chance: Rearranging the Unwanted Mass," *American Health* 12 (March 1993): 18.
66. Maria Paul, "Next Generation of Breast Implants Could Put Silicone and Saline to Shame," *Chicago Tribune*, September 27, 1998.
67. Susan Bell, "A New Model of Medical Technology Development: A Case Study of DES," *Research in the Sociology of Health Care* 4 (1986): 3.
68. Donna J. Haraway, "A Cyborg Manifesto: Science, Technology, and Socialist-Feminism in the Late Twentieth Century," in *Simians, Cyborgs, and Women: The Reinvention of Nature* (New York: Routledge, 1991).

Index

About the Author

Nora Jacobson is a medical sociologist who holds a Ph.D. from the Johns Hopkins University School of Hygiene and Public Health and an M.A. from the University's Writing Seminars. She was a postdoctoral fellow at the Mental Health Services Research Training Program at the University of Wisconsin, Madison. Her research focuses on the ways in which social constructions of health and illness affect the making of health policy.